Digital Transformation for the Process Industries

Digital Transformation for the Process Industries

A Roadmap

Osvaldo A. Bascur

Contributing Editor

Jim O'Rourke

CRC Press
Taylor & Francis Group
Boca Raton London New York

CRC Press is an imprint of the
Taylor & Francis Group, an **informa** business

MATLAB® is a trademark of The MathWorks, Inc. and is used with permission. The MathWorks does not warrant the accuracy of the text or exercises in this book. This book's use or discussion of MATLAB® software or related products does not constitute endorsement or sponsorship by The MathWorks of a particular pedagogical approach or particular use of the MATLAB® software.

First edition published 2021
by CRC Press
6000 Broken Sound Parkway NW, Suite 300, Boca Raton, FL 33487-2742

and by CRC Press
2 Park Square, Milton Park, Abingdon, Oxon, OX14 4RN

© 2021 Taylor & Francis Group, LLC

CRC Press is an imprint of Taylor & Francis Group, LLC

Reasonable efforts have been made to publish reliable data and information, but the author and publisher cannot assume responsibility for the validity of all materials or the consequences of their use. The authors and publishers have attempted to trace the copyright holders of all material reproduced in this publication and apologize to copyright holders if permission to publish in this form has not been obtained. If any copyright material has not been acknowledged please write and let us know so we may rectify in any future reprint.

Except as permitted under U.S. Copyright Law, no part of this book may be reprinted, reproduced, transmitted, or utilized in any form by any electronic, mechanical, or other means, now known or hereafter invented, including photocopying, microfilming, and recording, or in any information storage or retrieval system, without written permission from the publishers.

For permission to photocopy or use material electronically from this work, access www.copyright.com or contact the Copyright Clearance Center, Inc. (CCC), 222 Rosewood Drive, Danvers, MA 01923, 978-750-8400. For works that are not available on CCC please contact mpkbookspermissions@tandf.co.uk

Trademark notice: Product or corporate names may be trademarks or registered trademarks, and are used only for identification and explanation without intent to infringe.

For Product Safety Concerns and Information please contact our EU representative:
GPSR@taylorandfrancis.com
Taylor & Francis Verlag GmbH, Kaufingerstraße 24, 80331 München, Germany.

ISBN: 978-0-367-22237-6 (hbk)
ISBN: 978-0-367-53518-6 (pbk)
ISBN: 978-1-003-01052-4 (ebk)

Typeset in Palatino
by Lumina Datamatics Limited

To all my customers in their journey towards a digital transformation.

—Osvaldo A. Bascur

To my parents, the best ever.

—Jim O'Rourke

Contents

Foreword xv
Preface xix
Acknowledgments xxv
Authors xxvii
Commonly Used Terms and Abbreviations xxix

1. **Advancing to an Industrial Digital Data Infrastructure** 1
   The Disaster 1
   Journey to an Enterprise Industrial Digital Infrastructure 2
   Assessing the Current State 4
   Identifying the Problem 6
   The Current Data Infrastructure 9
   Understanding the Barriers to Success 12
      The Process Engineer's View 12
      The Plant Manager's View 15
      The Planning and Economics Coordinator's View 16
      The Production Manager's View 17
      The Maintenance Manager's View 20
      The IT Department's Role 21
      The Cost of Downtime 23
   Envisioning a Strategy 24
      Creating a Breakthrough Vision 25
      Assessing Data Infrastructure Maturity 26
      Laying a Foundation 31
         First Step to Action: Convening an Operational Team 31
         Engaging Corporate Leadership and Plant Employees 33
   What You Should Take Away 34
   References 35
   Additional Reading 35

2. **Building the Foundation** 37
   Chapter Overview 37
   The View from the Refinery Manager's Office 37
   Identifying Workflow Information Gaps 40
   Improving Scheduling, Operations, and Maintenance 41
   Process Unit Template for Smart Thinking 44
   Operational Excellence Methods and Tools 46
      Workflow Management 47
         The First Loop: Are We on Target? 48
         The Second Loop: Are We Satisfied? 49

Presenting the Digital Data Infrastructure Project
to the Entire Company 50
    Operations Standardization for an Enterprise Competence Center 52
        Enterprise Level 54
        Regional Level 54
        Individual Refineries 55
What You Should Take Away 56
References 57
Additional Reading 58

3. Using EIDI Data as a Strategic Asset 61
Chapter Overview 61
The Need for an Enterprise Industrial Data Infrastructure 61
Determining What Is Important 63
What Is an Enterprise Industrial Data Infrastructure? 65
A Decision Is Made 68
A Modern EIDI Implementation Approach 69
Decision-Making through Plant Data Hierarchy 71
EIDI Deployment and Configuring the EIDI Templates 73
    Using Templates to Define Assets and Event Frames 73
Data Object Models 75
Data Acquisition, Validation, and Classification 77
Block and Process Flow Diagrams 78
    Block Flow Diagrams 78
    Process Flow Diagrams 79
        Organizing Operational Data 81
        Process Flow Diagram Leads to Digitizing the Plant 82
What You Should Take Away 82
References 84
Additional Reading 85

4. Advanced Analysis Using Unit Data and Event Templates 87
Chapter Overview 87
Innovative Use of EIDI Capabilities 88
    Step 1: Develop Unit Process Templates Using Plant Block
    Diagrams 89
    Step 2: Analyzing and Visualizing Operational Variance from
    the Generated Events 93
        Classifying Asset Behavior for Process Improvement 96
    Step 3: Employing Offline Visualization Tools Using
    Unit Process Template Data 96
        Advanced Visual Analytics 96
        Using Microsoft Excel Analytics Tools 100

    Step 4: Using Contextualized Data for Modeling
    and Predictive Analysis 100
        Data-Driven Analytics 100
    What You Should Take Away 103
    References 105
    Additional Reading 106

5. **The Humans behind the Data: Visualization and Collaboration** 109
    Chapter Overview 109
    Who Will Be the ProcIndustries EIDI Users? 111
    The Impact of Change 112
    Rome Wasn't Built in a Day 113
    People-Driven Benefits from the EIDI Implementation 116
    Getting the Visualization Right 117
        Best Practices on Process Graphics and Ways to Present and
        Share Information 119
        Mobile Access to Information 120
        Asset-Relative Displays 121
        Process Improvement through Visualization 122
        Creating Dynamic Performance Operational Displays 124
    Data Turns into Workflows: Workflows Adopted by People Turn
    into Meaningful Change 126
    What You Should Take Away 127
    Reference 127
    Additional Reading 128

6. **Preventing Abnormal Situations** 129
    Chapter Overview 129
    Real-Time Data Analytics to Improve Operational Support 130
        Business Objectives Pyramid 131
            The Left Side of the Pyramid 131
            The Right Side of the Pyramid 133
        Enhancing Equipment Availability 135
            Reactive Maintenance 136
            Preventive Maintenance 136
            Condition-Based Maintenance 137
            Predictive Maintenance 137
            Condition-Based Maintenance: The P–F Curve 139
            Condition-Based Maintenance Using a Real-Time Data
            Infrastructure 141
                Assigning Context to the Equipment Template 142
                Assigning Analytics to Detect Abnormal Conditions and
                Trigger Events 143

        Generating Notifications from the Events 144
        Assigning the Event Templates for Analysis and Root
        Cause Determination 144
        Assigning Equipment Parameter Tables 144
    Pump Asset Example 146
        Pump Monitoring and Analysis 148
        Visualizing Pump Actionable Output 150
Integration with Other Systems 151
Implications of Improving Asset Availability 152
Operational Performance Management 154
    Enhancing Process Control Performance Monitoring 154
What You Should Take Away 157
References 159
Additional Reading 160

7. Energy Management and Operational Improvements 163
Chapter Overview 163
Revisiting Energy Consumption 163
    International Energy Standards 165
    Develop Clear Business Objectives 165
The Plan 167
    Metering and Inputs 168
    Data Capture and Reporting 169
    Data Analysis, Visualization, and Reporting 169
    A Takeaway for the Team on Measuring Power Consumption 173
    Smart Grid and Refinery Resiliency Improvements 173
Using Process Flow Diagrams for Energy Management 175
Using Advanced Analytic Tools to Gain Real-Time Insights 177
Mass Balances and Data Reconciliation 178
What You Should Take Away 179
References 180
Additional Reading 181

8. Successful Examples of Enterprise-Wide Digital Transformation 183
Chapter Overview 183
OilCo—Use of EIDI Data in Downstream Refining 184
    OilCo's EIDI Deployment 185
    Advanced Analytics and Machine Learning Leads to
    Significant Return on Investment 187
    On-Premise versus Cloud Analytics? No: On-Premise Plus
    Cloud 188
    OilCo's Return on Investment 189
    Lessons Learned 190

Use of EIDI Data by MidPetCo for Pipeline Operations 190
  Overview and Challenges 191
  MidPetCo 2.0 Initiative Leverages Operational Data for
  Profitability 191
  Real-Time Visibility Enhances Decision-Making and Reduces
  Inefficiencies 193
  MidPetCo's Return on Investment and Future Plans 194
GoldMineCo—Use of EIDI Data in Gold-Mining Operations 195
  Process Plant Optimization Successes at GoldMineCo's Mines
  and Plants 197
  Combining Time-Series Data and Location Data for Mobile
  Asset Maintenance 197
  Big Data Analytics—Populated by Site Information 198
  Other Initiatives to Extract Value from Data 198
  GoldMineCo's Use of the Digital Plant Template for Metallurgy
  Analysis 199
  Future Plans 199
  So, Was It Worth It? 200
MaterialsCo—Use of EIDI Data in Building Materials
Manufacturing 200
  Creating a Single Version of the Truth on a Global Scale 200
  Integration with Corporate Business Intelligence Software 202
  Forging Ahead—MaterialsCo Production Management
  Version 2.0 203
  Lessons Learned from Global Deployment 203
  Moving to the Next Level—Artificial Intelligence and
  Closed-Loop Control 204
EIDI Data Use by ChemCo for Specialty Chemical Manufacturing 205
  Using Technology to Combat a Decreasing Market 205
  Putting Data to Good Use for Improved Operations 206
  Reducing Emissions and Energy Consumption through
  Data Analysis 207
  Asset Management and Other Data-Driven Improvements 207
  Lessons Learned and Next Steps 208
Reference 209

9. **Beyond the Refinery—Connecting the Ecosystem** 211
Chapter Overview 211
New Ways of Sharing Data to External Partners for
Additional Value 212
PI Cloud Connect 215
PI Cloud Connect Customer Examples 217

    Anglo American Platinum—Outotec 217
        Benefits for Anglo American Platinum 219
        Benefits for Outotec 220
    Cloud-Based Services Provided to a Power Generation
    Company 220
    PI Cloud Connect Scalability 222
    What You Should Take Away 223
    References 224
    Additional Reading 224

10. **Operational and Business Analytics Integration** 225
    Chapter Overview 225
    From Operational Intelligence to Enterprise Intelligence 226
    Types of Advanced Analytics 227
    Holistic Analytics to Meet Business Objectives 228
    ProcIndustries Integrates Operational and Business Data 230
    Integration to Corporate Analytics Systems 232
        Using a Real-Time Data Infrastructure versus a Data Lake
        Approach 233
    Using a Standard Method for Advanced Analytics Integration 235
    Integrating Production Event-Based Data to Advanced Analytics 235
    Creating a Digital Twin for Process Simulation and Modeling 239
    Integrating Time-Series Data with Geospatial Systems 240
    What You Should Take Away 242
    References 243
    Additional Reading 243

11. **ProcIndustries Enterprise-Wide Rollout** 245
    Chapter Overview 245
    ProcIndustries Continues Their Digital Transformation 245
    Building the Business Case 247
        Financially Justifying the Proposed EIDI Enterprise Rollout 248
        Quantifiable Benefits 250
        Non-Quantifiable Benefits 252
        Management's Decision 253
    Deployment of the Infrastructure and Initial Use Cases 255
        Working Toward a Smooth EIDI Enterprise-Wide Rollout 256
        Initial Rollout Activities 258
        EIDI Architecture and Cybersecurity 260
    Phase III: Ongoing Governance and Future Use Cases 260
        Future Use Cases 261
    What You Should Take Away 262
    References 263
    Additional Reading 264

Contents   xiii

12. The Future of the Digital Enterprise 265
   Chapter Overview 265
   How the Most Successful Companies Achieve Success 265
      Alignment of Business Goals with Digital Strategies 266
      Viewing Operations and Production Data as a Critical
      Strategic Asset 267
      Successful Teaming with Strategic Partners 267
      Architecting for the Unexpected 268
      Effective Use of Analytics 268
      Self-Serve Access to Needed Information 269
   ProcIndustries Enterprise-Wide Rollout Results 269
   Cloud Strategy 270
   Edge Analytics and the Evolution of the Control System 271
   The Leadership Meeting 273
   Hybrid Cloud Strategy 275
   The Next Challenge 276
   And the ProcIndustries' Story Continues 278
   What You Should Take Away 279
   Additional Reading 279

Index 281

Foreword

I've had the distinct pleasure of working with Osvaldo A. Bascur for more than 21 years. I first met Osvaldo as a new OSIsoft hire in 1998. Since then, I have seen him provide insight and guidance to customers in many different industries, particularly mining and metals and downstream refining. His thought leadership has helped people around the globe to better understand and use technology to achieve their business objectives.

His vision of capturing and analyzing operations information is unparalleled. He's worked with the brightest minds in the process and metals industries to have them derive the most value from their data. In this book, he shares his data analysis secrets and shows how you can implement the same thing he has done for our customers. It's been a privilege to work with Osvaldo on this book, and we hope that the ideas presented here will be a roadmap for your company to master your business and digital transformation.

How This Book Will Assist You

The process starts when you realize the incredible value of your operating data. Osvaldo and I have seen enterprises achieve some astonishing accomplishments when they use their operating data effectively and transform their business into something they hadn't imagined.

For example, when pulp and paper manufacturers experienced reduced demand during the 2000s, a few imaginative employees discovered they could sell the excess electricity that they made back to the local power company. Not only did they use their operating data to make their own operations cost-effective, they uncovered a brand new revenue source.

Another example is when a former colleague of ours wrote a calculation that monitored the health of 3,800 electronic submersible pumps (ESPs) in real time. In the first year, the company increased ESP uptime by 63%, resulting in a first year savings of US$40 million. This savings was derived from spare parts savings and avoided maintenance labor to service the failed ESPs. The most striking takeaway is that the US$40 million savings did not include lost production revenue at a time when a barrel of oil cost more than US$100. To watch Industrial Evolution's 20-minute presentation starting at minute 10:30, see www.osisoft.com/poc.

There are many success stories such as the two just mentioned. But how do people know where to start? Osvaldo and I put together this book to provide readers a blueprint and step-by-step plan to digitally transform plant facilities and entire enterprises. You will find out how to begin and how to define the best use cases to transform your data into tangible business value, with tangible financial results.

If you are new to this type of real-time data infrastructure software, you will learn how industry uses this software to achieve substantial business returns by stemming inefficiencies; enhancing work processes; and improving collaboration within the company and with partners, suppliers, and customers.

If you are experienced using this technology, we think you will find some new use cases to find business value for your company or for your clients. In Chapter 4, you will find a link to help you implement contextualization of operations data into significant event data, and how to use this data for advanced analytics, such as machine learning (ML) and big data analytics.

If you are somewhere in between, we think the use cases will inspire you to use and reuse your data in ways you hadn't imagined. We have attempted to provide enough specifics so that people and companies have enough information to strategize and begin new initiatives. At the same time, we have tried to steer the book toward a conceptual approach, rather than a "how-to" implementation manual. There are several links throughout the book that offer more detailed information on some of the topics we address, such as best practices for condition-based maintenance or how to construct a smart unit template to capture and analyze events.

Once a real-time software information system (referred to as an *enterprise industrial data infrastructure* or *EIDI* in this book) is deployed as the time-series *system of record*, the magic happens. Businesses achieve the following realizations:

- The sensor data captured from plant/refinery control systems becomes contextualized and transformed into valuable information, to be used by operations, engineering, and business personnel alike, bridging these worlds.
- The real-time and historical data can be reused many times for numerous use cases, such as asset health, process improvement, root cause analysis, regulatory compliance, and as input to big data analytics.
- The company can transform into an efficient, data-driven culture, enhancing profitability, customer satisfaction, and environmental stewardship.

Some important advice we can offer is not to think of an EIDI only in the context of a data historian. While this is a major component of an EIDI, you can achieve much more value if you deploy it as the real-time lifeblood of your operations. To only think of it as a data repository for forensic analysis and reporting is, in our view, a bit shortsighted. In this book, you will find the best practices we have seen from our brightest and most innovative clients.

Although this book uses the fictional ProcIndustries to step readers through how a refinery realizes their goal of becoming a digital enterprise, the process they undertake, the issues they solve, and the benefits they achieve are 100% real. Make no mistake about that. These results reflect what both Osvaldo and I have seen our clients accomplish time and again. Each chapter addresses a topic or business use case that transcends specific companies or industries. Many of these issues are the same whether you are refining oil, generating power, or moving natural gas across the country.

Please enjoy the journey of ProcIndustries and their South Texas refinery. We hope your takeaways and outcomes are as fruitful as the ones described in this book. If you have any questions for us regarding digital transformation strategies or our software, please contact us at digitaltransformationbook@osisoft.com.

Jim O'Rourke

Contributing Editor

Preface

Peter and ProcIndustries—A Story for Every Industry

This is the story of ProcIndustries, a fictional, midsized, downstream oil-refining company. The book describes how the company navigates its digital transformation journey. In this book, you may think of a digital transformation as the changes associated with applying, embracing, and achieving successes using digital technology.

The company has been operating for 35 years and is a recognized industry leader, but times are changing quickly. Global competition is fierce, energy costs have become cyclical, and the price and availability of raw materials are more volatile than ever before. ProcIndustries' established workforce is aging, with many retiring as they turn 60 years of age. New workers enter the company with new skills and different expectations. Deregulation is paradoxically coupled with stricter safety and environmental compliance laws. Dramatic advances in technology present great opportunities for operational improvements and increased profitability. With opportunities, there are often great challenges to overcome, such as the following:

- Where do we begin?
- What problems should we tackle first?
- How do we decide what is the best technical fit for us?
- Do we buy a software infrastructure or a solution tailored to our company?
- How will we justify purchasing it?
- What is the cost of not doing anything?
- What results should we reasonably expect to realize from deploying it?
- Will our people support it and will they be equipped to successfully utilize it?

To gain clarity, ProcIndustries hires Peter Argus, an operations expert who will attempt to guide ProcIndustries through this decision and navigate positive outcomes. As Peter steps into his new role, he recognizes that he must help ProcIndustries adapt to new realities. It is his mission to transform the company from a collection of silos to an enterprise with connected functions that empower everyone—from C-suite executives to frontline managers—to make data-driven decisions using real-time intelligence. Peter recognizes that effective data consumption and willing collaboration are the keys to transformation.

In sharing Peter's experiences, this *Digital Transformation for the Process Industries: A Roadmap* navigates the reader through continuous process improvements, including status assessments, strategy design, and the adoption of industrial digital data infrastructure best practices. Through Peter's eyes, we see ProcIndustries' strengths, areas for improvement, human assets, process units, and challenges as he undertakes the mission of remaking his company; however, for Peter to implement data-driven processes and deliver business insight, he must overcome entrenched and outdated business practices.

His journey begins with assessing the current state of ProcIndustries' operations and building consensus among managers and executives that a data-centric infrastructure will lead to improvements. It quickly becomes clear that it will take more than information technology to deliver substantive business benefits. Business processes will have to fundamentally change. People will need to adapt their working styles and tasks. And it will take commitment, time, and investment to make sure that all the players are speaking the same language.

The story of Peter and the staff from leadership to maintenance at ProcIndustries may be fictional, but their experiences reflect real-world challenges faced by leaders and managers across industries, including

- Dealing with competition in a global economy that has seen dramatic changes in resource supplies and demand, regulatory protocols, and labor demographics;
- Implementing the cultural and technological changes required to provide real-time information—and then empowering people to identify and take action;
- Managing and analyzing oceans of data flowing into and through a company—and developing a common language for the organization to take advantage of it; and
- Training people to do their jobs differently as new tools enable them to see business conditions in new ways and take appropriate actions as a result.

While the story focuses on one industry, this book can benefit executives and managers in every industry. Every industry needs organizational expertise, company-wide training, implementation plans, change-management processes, and most importantly, the cultural change necessary to empower personnel from frontline workers to management to effectively use this real-time information.

The ability to capture and translate data into meaningful information—making it possible to assign the right priority to any given event—is crucial in every industry. In retail, stores must train in-store employees to make decisions at the point of sale to ensure customer satisfaction and deepen brand loyalty. In aviation, the failure of an airliner turbine can lead to wasted fuel, scheduling delays and missed flight connections, emergency landings, or, at worst, catastrophic events. In mining, enterprises must monitor equipment fleets in real time to assess maintenance needs and prevent equipment failures that lead to costly downtimes. Well-managed equipment fleets result in productivity gains and energy and water cost savings, and correlate with fewer safety incidents and stronger environmental compliance.

It is our hope that over the course of this story, Peter shows you as well as ProcIndustries how to simplify the semantics of process industries, how to empower frontline employees, how to detect and prevent potentially adverse events, and how to reduce costs with reliable business intelligence. And we hope that by the story's end, you, like ProcIndustries, will emerge empowered to provide the right data context for everyone in your organization, so that computer systems can optimize human workflows for collaborative decision-making, and innovative technologies can catalyze a culture of continuous improvement.

The Transformation Journey

Chapter 1, "Advancing to an Industrial Digital Data Infrastructure," sets the stage for the ProcIndustries' story. After a transformer accident disrupts refinery operations, management asks why such incidents cause widespread chaos and how these events can be avoided in the future. The current state of the company is assessed, and the problems are identified through a series of information-gathering meetings with key players. We see that a digital transformation needs to be aligned with the company mission and implemented as a continuous program rather than a finite project or application. The notion of an operational excellence program is put forward, with operational and safety issues at the forefront. This chapter ends with a three-step deployment plan approved by management.

In Chapter 2, "Building the Foundation," the ProcIndustries South Texas refinery digital transformation team is formed. In this chapter, readers learn how the team goes about prioritizing initiatives to reduce the problems that have been plaguing the refinery. They must also streamline workflow across the refinery teams so that everyone has access to the data they need in a way that will allow them to make better decisions.

In Chapter 3, "Using EIDI Data as a Strategic Asset," the first milestone is reached, as the ProcIndustries team selects the technology they will deploy as their enterprise information data infrastructure (EIDI). An overview of the software is presented in the context of how key functionality can be used to achieve the refinery's strategic business objectives. Also explained is how EIDI data is transformed from raw sensor values to important contextualized information.

In Chapter 4, "Advanced Analysis Using Unit Data and Event Templates," the implementation team begins to make decisions as to how the EIDI will be deployed so that initial returns are quickly realized. At the same time, the team decides that standard, reusable templates to classify and contextualize the data will reap benefits during the initial pilot phase and throughout the life of the technology, as they must also plan for future use cases. This chapter presents a strategy for an accelerated adoption using standard unit object models. The team decides that information collected and analyzed will be most useful when classified by specific production events. This chapter will appeal to advanced readers and contains web links to supporting material.

No matter how impactful the technology, humans are responsible for whether a software project is successful. Careful thought must be given to the ways information is provided to them and whether they are empowered to make decisions that impact company operations. In Chapter 5, "The Humans behind the Data: Visualization and Collaboration," readers discover ProcIndustries' plans to ensure that team members collaborate effectively to solve business problems at hand.

To achieve optimum effectiveness, the team designs and implements a strategy where the EIDI continuously monitors refinery equipment and the process parameters. When an unexpected situation occurs, the workers are notified right away, so that decisions can be made at the earliest possible time. Chapter 6, "Preventing Abnormal Situations," introduces the concepts of condition-based and predictive asset monitoring for best practices in asset management.

The South Texas refinery soon realizes initial gains from their EIDI deployment. With equipment and processes running smoothly with a minimum of unscheduled downtimes, the digital transformation team focuses its efforts on reducing energy costs. In Chapter 7, "Energy Management and Operational Improvements," readers learn how the team's innovative design strategy allows them to analyze energy consumption in ways they never had been able to visualize.

In Chapter 8, "Successful Examples of Enterprise-Wide Digital Transformation," readers discover how large energy, process, and metals companies have implemented their own continuous improvement. Readers learn how these companies in various industries used EIDI standardization to simplify their implementation, maintenance, and enterprise-wide analysis of process operations and their fleet of assets. We present use cases from actual companies, using an EIDI in innovative ways to become industry leaders. These include an international integrated oil and gas company, a North American midstream petroleum services company, an international gold-mining company, a multinational building materials producer, and an international specialty chemicals company.

Preface xxiii

Now that results are being achieved inside the South Texas refinery, the ProcIndustries team becomes aware that the data they have been collecting can be valuable for solving ongoing problems where outside expertise is needed. They work remotely with their equipment suppliers, supply chain vendors, and software partners to accomplish this.

In Chapter 9, "Beyond the Refinery—Connecting the Ecosystem," the digital transformation team finds a way to securely transmit their data to these companies using secure, cloud-based technology. Relevant use cases from an international precious metals producer and a water-treatment chemicals company are presented.

Successes at the South Texas refinery are becoming known throughout the company. ProcIndustries' data scientists and information technology (IT) personnel have suggested that the refinery data be utilized by corporate analytics to further extend use of the information the EIDI collects and stores. Chapter 10, "Operational and Business Analytics Integration," shares how the digital transformation team goes about extracting and exposing plant data for corporate analytics, models, and machine-learning algorithms. Best practices are highlighted, where leading companies combine time-series operations data with other types of data for powerful offline analytics.

Now that the South Texas refinery team has achieved substantial results from implementing EIDI at their location, the company contemplates scaling the technology to their other refineries and plants to maximize fleet-wide process efficiency and to ensure that all assets are operating at peak availability. Management will now have real-time situational awareness to effectively assess fleet performance and where to effectively spend capital.

In Chapter 11, "ProcIndustries Enterprise-Wide Rollout," readers discover the justification and approval process for an enterprise-wide deployment, as well as best practices for accelerated implementation and adoption, emulating the South Texas refinery methodology. By the end of the story, ProcIndustries has become an example of operational excellence. Real-time, enterprise-wide visibility, data-driven processes, and insightful analytic and visualization tools have put the company on a path to sustained profitability and innovation.

Chapter 12, "The Future of the Digital Enterprise," explores the challenges of moving to Industry 4.0—embracing and integrating emerging digital technologies and cloud-based strategies. Companies like ProcIndustries have many sites that have a need for their data: remote sites, cloud-based vendors and partners, and the community at large. The chapter summarizes the new ways people work and collaborate, the new business processes, and new ways of using information. The chapter concludes by sharing how Peter Argus and the ProcIndustries digital transformation team navigate new and unplanned journeys.

MATLAB® is a registered trademark of The MathWorks, Inc. For product information, please contact:

The MathWorks, Inc.
3 Apple Hill Drive
Natick, MA 01760-2098 USA
Tel: 508 647 7000
Fax: 508-647-7001
E-mail: info@mathworks.com
Web: www.mathworks.com

Acknowledgments

This book began in conference lectures, technical papers, and chapters in books on industrial technology. I am truly indebted to all my clients who contributed at the inception and testing of the concepts and methods. Many original seminal papers written in collaboration with my mentor Dr. J. Pat Kennedy have been adapted to our Digital Age. The early vision of having a data infrastructure as a part of a living plant and organization is a fundamental concept that many take for granted.

I thank everyone who discussed the subject with me and made useful suggestions. I am greatly indebted to Keith Pierce, Jim O'Rourke, Mariana Sandin, and Ales Soudek for their careful reading of the manuscript and offering corrections and suggestions for improving the text. Ales, Mariana, and I started many years ago when we felt that our technology could transform raw data into inFORMation by adding the right time context with a top-down approach. Special thanks to Jim, who acted as Contributing Editor, and his contribution during the second phase of the creation of the book.

Other colleagues supplied advice and concept improvements and, most importantly, encouragement to continue when I was struggling. Great thanks are given to the wonderful Gina Laviste for turning my rough sketches into beautiful figures. Much must be done to turn the manuscript into a polished book, and I wish to thank Terese Platten for her meticulous editorial work and for her patience in dealing with an inexperienced author. The title of the book evolved from suggestions made by Dr. Pat Kennedy and Dr. Dominic John.

The book was made possible by the long-term vision of OSIsoft to allow me to write the book and work very closely with many clients to test and implement these concepts and methods. The first draft was done in August 2016 and put on hold. During these last years, I had the opportunity to work with customers to validate the concepts and methods. In particular, I would like to thank Ted Gorrie, Dave Draper, Daniel Gervais, Gio Chisotti, Ann Moore, Michael Halhead, Michel Plourde, Stephane Paquet, Barun Gorain, Jaco Steyn, Norm Doucet, Andrew Cooper, Rob Dunne, and Neelesh Chawda for many conversations and their leadership and feedback. This book would not be possible if I did not have the industrial validation and encouragement from many clients and colleagues.

Because much of this material has been presented in universities for students to become familiar with time-series data and event-framing data aggregation at the desired degree of detail, I would like to thank John Matranga for his encouragement to create and deliver these lectures in academia and industrial conferences. The students and practitioners can perform their own simulations and experiments, and after earning their engineering degrees, students are prepared to integrate data science, business intelligence, and machine learning tools with the industrial world. And finally, many thanks to my family and close friends; without their unconditional love, this book would not exist.

Osvaldo A. Bascur
The Woodlands, Texas

I'd personally like to thank everyone that helped me create content or had a part in reviewing this book. First, Dr. Osvaldo A. Bascur for his unique insights on time-to-value and his invaluable contributions to this book and to OSIsoft customers, Curt Hertler for advanced analytics, Keith Pierce for his thought leadership regarding best practices in condition-based and predictive maintenance, John Matranga and Richard Beeson for their leadership help in peering into the future, and Dr. Pat Kennedy for creating the technology and the company that makes this book possible.

Two contributors went above and beyond in sharing their personal intellectual property acquired over many years of thought leadership and being a resource to our clients. Craig Harclerode distilled the ingredients of what makes a company successful in navigating a successful digital transformation. Allon Shiff was instrumental in identifying quantifiable value that our customers achieve by following the blueprint described in the book.

My thanks also go out to Terese Platten, who helped us tremendously in creating something that we hope is beneficial and enjoyable to read.

Finally, I'd like to acknowledge our process industries and energy provider customers, who inspired many of the ideas in this book by their continuous innovation and applying the technology in ways we never could imagine.

Jim O'Rourke
Houston, Texas

Authors

Osvaldo A. Bascur received his BS in chemical engineering and BS in metallurgical engineering from the University of Concepcion, Chile and a PhD in metallurgical engineering from the University of Utah, Salt Lake City. He was a process control engineer working for Duval Corporation, Tucson, Arizona (now Freeport-McMoRan). He was transferred to Pennzoil Company in Houston, Texas, where he worked as a staff engineer in the process-engineering department. He was a principal for more than 25 years at OSIsoft LLC. He is currently Principal Digital Transformation for OSB Digital, LLC and Seeq Advanced Analytics consultant. He received the Society for Mining, Metallurgy & Exploration (SME) 2013 A. Gaudin Award. His current focus is the integration of sensor data for operating intelligence and asset and energy optimization. His work includes consulting for the process industries and preparing academic lectures in process analysis and predictive analytics. Dr. Bascur is a member of SME, Association for Iron and Steel Technology (AIST), American Institute of Chemical Engineers (AIChE), and International Federation of Automatic Control (IFAC) Mining, Mineral and Metal Processing (MMM).

Jim O'Rourke is an academic principal with OSIsoft LLC residing in Houston, Texas. He is responsible for OSIsoft's strategic relationships with many universities in the United States. Jim also identifies and develops key alliances with commercial companies and consortia in the academic space. Jim has been with OSIsoft for 22 years. Prior to his current academic role, he was responsible for power and utility sales and business development in Texas; Sales Director in Russia and the CIS States for two years; and U.S. Federal Partner Manager for three years. Prior to joining OSIsoft, Jim worked at Biles & Associates, first in software development and project deployment, then in sales and business development. He is a graduate of New Jersey Institute of Technology.

Commonly Used Terms and Abbreviations

Commonly Used Terms

OSIsoft Products

PI Asset Framework: Allows the definition of consistent representations of organizational assets and/or equipment so users can define relationships between their data to make information more understandable, add metadata and context to information, and ensure data consistency across the enterprise.

PI Cloud Connect: Cloud-based mechanism for securely sending and receiving data from one PI Server, typically a production company or manufacturer, across corporate boundaries, to another company's PI Server, typically a service provider, vendor, or solution provider.

PI Event Frames: Organizes production data stored in the PI System data archive by critical events, such as downtimes, excursions, startups, shut downs, curtailments, product batches, paper grades, and shifts to simultaneously analyze and compare data related to user events in context.

PI Integrator for Business Analytics: Combines contextualized PI System data with non-OSIsoft business intelligence tools for retrospective analyses on real-time PI System data.

PI Integrator for Esri ArcGIS: Combines PI System data with Esri's ArcGIS geospatial information system (GIS), so users can combine and analyze sensor-based, time-series data from the PI Server within the geospatial context of Esri maps.

PI Integrator for SAP HANA: Combines PI System data with the SAP high-performance analytic appliance (HANA) to analyze sensor-based, time-series data from the PI Server with the high-performance analytics and database of SAP HANA.

PI System Explorer: Sometimes referred to as PSE or AF Client; the PI AF user interface, allowing users to find information about their equipment and processes. PSE also is used to configure and manage PI AF, PI Notifications, and PI Event Frames.

xxix

PI Vision: Browser-based software to analyze and visualize PI System data across mobile devices or personal computers. PI Vision supports PI Event Frame analysis, PI Asset Framework unit relative displays, and other integrated PI System features without the use of programming.

Microsoft Products

Azure Machine Learning Studio: A machine learning service, which provides a centralized place for data scientists and developers to work with artifacts for building, training, and deploying machine learning models.

Cortana: A digital assistant program.

Excel: Spreadsheet software found in Microsoft Office or Office 365.

Flow: Now called Power Automate; cloud-based software that allows users to create and automate workflows and tasks across multiple applications and services without help from developers. Automated workflows are called *flows*. To create a flow, the user specifies what action should take place when a specific event occurs.

Power BI: Data and analytics reporting tool that helps organizations bring together disparate data sets into reporting dashboards.

PowerApps: A suite of applications, services, connectors, and data platform that provides a rapid application development environment to build custom applications for business needs.

Amazon Products

Amazon QuickSight: Business intelligence tool under the umbrella of the Amazon Web Services (AWS) platform; provides visualizations, interactive dashboards, and machine learning insights.

Amazon Web Services: Subsidiary of Amazon that provides on-demand cloud computing platforms.

Other Terms

asset object model

big data

block flow diagram

condition-based monitoring

control loop

digital age

digital data infrastructure
event frame
operational states: running OK; in trouble; idle; down; in maintenance
process flow diagram
time-event-framed template
time-slice

Abbreviations

AAP	Anglo American Platinum
AF	Asset Framework
AI	artificial intelligence
APR	advanced pattern recognition
AWS	Amazon Web Services
B2B	business-to-business
BEP	best efficiency point
BI	business intelligence
CAPEX	capital expenditure
CBM	condition-based maintenance
CII	continuous-improvement and innovations
CIO	chief information officer
CMMS	computerized maintenance management system
CSM	customer success manager
CTO	chief transformation officer
CUSUM	cumulative sum
DCS	distributed control system
DCU	delayed coking unit
DMAIC	define, measure, analyze, improve, and control
DMZ	demilitarized zone
EA	enterprise agreement
EBITDA	earnings measured before deductions for taxes, interest expenses, depreciation, and amoritization
EIDI	enterprise information data infrastructure
EPA	U.S. Environmental Protection Agency

ERP	enterprise resource planning
EU	European Union
EWMA	exponentially weighted moving average
FCCU	fluid catalytic cracking unit
GIS	geospatial information system
HA	high availability
HAZOP	hazard and operability
HVAC	heating, ventilation, and air-conditioning
ICC	Integrated Collaboration Center
IIoT	industrial Internet of things
IoT	Internet of things
ISO	International Organization for Standardization
IT	information technology
KPI	key performance indicator
LIMS	laboratory information management system
LP	linear programming
ML	machine learning
MSPC	multivariable statistical process control chart
NDP	New Downstream Program
NGL	natural gas liquids
NPSH	net positive suction head
ODBC	open database connectivity
OPE	overall production effectiveness
OPEX	operational/operating expenditure(s)
OSHA	Occupational Safety and Health Administration
OT	operational technology
P&ID	piping and instrumentation diagram
PaaS	platform as a service
PCA	principal component analysis
PFCD	process flow control diagram
PI AF	PI Asset Framework
PI CC	PI Cloud Connect
PIMS	Plant Information Management System
PLC	programmable logic controller
PLS	projection of latent squares

PM	preventive maintenance
R&D	research and design
ROCE	return on capital employed
ROI	return on investment
R_xM	prescriptive maintenance
SCADA	supervisory control and data acquisition
SME	subject matter expert
SPC	statistical process control
SQC	statistical quality control
TA	technical advisor
UPS	uninterruptable power supply
VOC	volatile organic compound
VPN	virtual private network
VR	virtual reality

1
Advancing to an Industrial Digital Data Infrastructure

In God we trust; all others must bring data.

W. Edwards Deming

The Disaster

It was a late-May gray day in South Texas. During this season, severe thunderstorms can disrupt the running of large industrial complexes. In this case, lightning hit an electrical transformer at the ProcIndustries oil refinery, causing it to overload and shut down one of their crude units.

That afternoon, Peter Argus was driving to his new job as a continuous-improvement manager at ProcIndustries when he heard the news on the radio: There had been an explosion at the company's refinery.

The news report stated that an electrical transformer had exploded and caught fire, causing the refinery's Crude Unit A to shut down. Peter knew that the refinery had two crude units. As Peter listened, he wondered about the ramifications for the people who worked at the refinery, and he decided to drive his pickup truck to the refinery rather than to the main office so he could observe the situation.

Peter wondered how prepared ProcIndustries was to handle this type of emergency. If it was just Crude Unit A affected as the radio news said, what would that mean? Was anyone hurt? What were the production losses? What was the environmental impact? Was the refinery workplace safe?

From his 15 years in the industry, including previous jobs in a power-generation facility and at a pulp and paper manufacturer, Peter knew that transformers are very stable. But like any piece of equipment, if a transformer is not maintained properly or is pushed to its limits, it will have problems and eventually break.

Peter parked far from the place where several crew members, firefighters, and a safety team were assessing the situation. He made sure his employee ID badge was showing as he approached. Peter saw many people were walking around to assess the damage. A man who Peter guessed was the plant manager instructed the crew to check for any process unit leaks, production losses, or environmental hazards caused by the lack of power (see the box "About the Process Unit"). The plant manager then talked to the authorities and reported that the fire was contained very quickly with no apparent injuries. While one production line was not operating, the plant manager said that preliminary checks at the refinery showed there did not appear to be any environmental damage.

> **ABOUT THE PROCESS UNIT**
>
> In this book, *process unit* is a term used to describe the area in a plant hosting the equipment required to transform raw materials into valuable semi-finished and finished products.

Peter was impressed that the plant manager appeared to have gained control of the situation and was working with emergency responders to handle this incident and its immediate aftermath. He wondered how the plant manager was able to find out information about the status of systems at the refinery and report it to the authorities at the scene.

As Peter climbed back into his pickup, he knew the incident would have ramifications for his new job at ProcIndustries. He was about to find out just how big those ramifications would be for his career and for the future of his new employer.

Journey to an Enterprise Industrial Digital Infrastructure

In this book, the character of Peter Argus represents managers at companies in a range of industries. They work with systems that track operation-related activities. However, they have a major challenge: They are not getting the information they need, when they need it, to make better decisions and improvements in their processes. As a result, their results are not as strong as they should be.

To survive, companies must create what Arie de Geus, the group planning coordinator for Royal Dutch Shell, called "the living company" (De Geus 1988). He theorized that the only sustainable competitive advantage is an organization's capability to learn faster than its competition. De Geus developed these strategies in the 1980s while looking for the secrets of corporate longevity and studying companies that had survived more than 100 years. He concluded that a company that involves everyone needed for execution in the planning and decision-making process will be more successful in a world which it does not control. De Geus concluded that it is not efficiency, but flexibility to adapt that an organization needs to survive. Companies must have methods, vision, and faith to institute a system that enables the existing staff to be flexible. People can alter the direction of the company with many small moves within and outside their domain (Kennedy 2002).

ProcIndustries is embarking on an operational excellence program. The company's leaders are betting on the future: the digitization of their physical world through advancing technologies such as the industrial Internet of things (IoT). As such, by reevaluating what they have and building a digital data infrastructure, they will apply smart thinking, engineering, and analytic tools to proactively improve awareness and avoid operational losses. They will enable their people to succeed.

Of course, this kind of operation is not easy to achieve. At its root, the challenge may be that they are not using their existing plant information systems to their full capabilities. It may be that they need to update their current business performance methods to take advantage of new sensors and data analysis capabilities that provide real-time insights into operations (see the box "Managers at a Disadvantage without Real-Time Data"). Or it could be a combination of these two factors.

MANAGERS AT A DISADVANTAGE WITHOUT REAL-TIME DATA

Many enterprises face competitive challenges, volatility in their raw material supplies, and changes in work processes and regulatory regimes. The collection and analysis of detailed data about operations can empower frontline managers and workers at every level, helping them manage these challenges. A digital data infrastructure enables this capability and brings real-time visibility into operations to every level of the company, from frontline workers to their managers to the executive suite.

Although the exact factors may differ, the voyage to improvement has common traits and a common destination: an enterprise industrial data infrastructure (EIDI) that brings insights to plant teams so they can make better decisions to improve operations. With EIDI, one plant can share lessons with others, and the larger enterprise can collect these lessons and develop best practices that the organization can codify and build on for a more sustainable and profitable future.

That is the goal for Peter and his colleagues at ProcIndustries. There are four stages to the journey:

1. **See the future.** Plant teams need to recognize the potential and value of an EIDI and articulate their vision.
2. **Gain management support and form the team.** Achieving a successful EIDI means winning support from stakeholders to make the vision a reality. Creating a team of stakeholders to lead the EIDI implementation is an essential part of that effort.
3. **Understand the barriers to success.** Only through a detailed understanding of existing practices can an EIDI be implemented. This calls for learning about how work is accomplished, the roles in a plant, and how each team member uses data and shares data. It requires interviews and information exchange.
4. **Proceed.** Armed with stakeholder support, an understanding of current conditions, and a vision of the future, the team leading an EIDI effort can move forward to implement the initiative.

In this book, as in real life, some of these activities can overlap and occur simultaneously. For example, support for an EIDI project among some stakeholders can build as they begin to understand the company's challenge and how an EIDI can help them. The journey itself can be hard work, but the destination is rewarding, not only for the team members who gain new skills but for the company as a whole.

Assessing the Current State

The next morning, Peter had a meeting scheduled at the South Texas refinery's administration building with Tom Jordan, the plant manager. While driving to the office, Peter considered what he'd already learned about his company.

ProcIndustries is a downstream oil company that makes gasoline, diesel fuel, and other distillates and coke for the U.S. and international markets. For many years, ProcIndustries had been content to be a successful midsized independent refinery. During that time, the company had seen a dramatic rise in competition from overseas players as well as significant variability of the crude oil supply. This variability negatively impacted the refinery's processing units, ultimately affecting run-time length and processing costs. In the past few years, they built several petrochemical plants to extend their value chain.

During interviews for his new job, Peter received a briefing on these trends and learned five pressing issues.

1. **Changing raw materials, such as crude oil.** The incoming available crude supplies show increasing contaminants, such as salts and solids. This trend has ramifications for production. Peter learned, for example, that ProcIndustries has seen a significant deterioration in its ability to adequately process crude oil laden with contaminants. In response, ProcIndustries altered its production equipment at the South Texas refinery. Although some of the problems were partially mitigated, they remained time-consuming issues for engineers, operators, and the company's chemical vendor to manage.

   Because processing has become more complicated, it requires better communications between the planning, operations, and process engineering teams. It also requires additional and more frequent quality tests and improved equipment sensors.

2. **Throughput limitations and rising energy requirements.** The accumulation of unwanted materials, called *fouling*, is caused by an increase of contaminants and other changes in the properties of oil and can lead to throughput limitations. In addition, fouling can contribute to rising energy demand in the refinery process because of the significant reduction in the heat transfer rates required for production.

3. **Fluctuating energy costs.** Besides the challenge presented by the variability in raw materials, fluctuating energy costs have strained the electrical grid feeding the company's four oil refineries. Recently, the company received requests to participate in smart grid initiatives to avoid problems with electric power distribution.

4. **Environmental regulations.** More stringent environmental regulations are putting significant pressure on ProcIndustries to ensure a virtual 100% compliance level on water reuse and water discharge requirements. For example, the refinery must have total control of benzene and other contaminants in water discharge, something mainly driven by the quality of raw materials being processed.

5. **Safety regulations.** ProcIndustries is also required to update its process safety management systems to comply with the new U.S. Occupational Safety and Health Administration (OSHA) rules and related regulations in other countries.

Identifying the Problem

The pressing issues that Peter learned about had hurt ProcIndustries' profitability. He knew that ProcIndustries' executive leadership, including the CEO, Jeff Edgell, and Bill Roberts, the vice president of operations (and Peter's boss), were concerned with the trajectory of the publicly held company's results. Peter learned during his interviews that the board of directors was demanding improvements.

Peter also knew he was hired because Bill was a champion for change at ProcIndustries and committed to bringing what he called *operational excellence* to the company. Bill had hired a top management consulting firm to assess the company's operations. ProcIndustries had in place a plant information system, but it was aging and not well used, and the consulting firm had recommended that ProcIndustries implement an EIDI program to replace it. This digital data infrastructure would become the system of record and enable the collection and analysis of data and the sharing of insights throughout the organization (see the box "What Is a Digital Data Infrastructure?").

WHAT IS A DIGITAL DATA INFRASTRUCTURE?

Historically, economists and engineers used the word *infrastructure* to describe the basic physical structure needed to support the operation of a society or enterprise, including roads, bridges, water supplies, sewer lines, electrical grids, and telecommunications. In recent years, the term infrastructure has also begun to encompass organizational concepts that support the overall health of an enterprise. A *software data infrastructure* that collects and distributes operational data and information is one of the most recent examples of this latter concept.

For many industries, deriving insight from large amounts of data to improve process efficiency, asset health, and product quality can be a lever of competitive differentiation. A digital data infrastructure acts as an information highway that parallels the physical infrastructure. It carries topologically mapped signals from machines, sensors, and external data sources to individual users, centers of excellence, and centralized diagnostic centers. These users share common data visualizations for collaborative analysis among people from a range of disciplines.

A digital data infrastructure creates an economy of scale by reducing the complexity and cost of storing decades of high-fidelity data. Through enhanced connectivity, an infrastructure ensures that people have the information they need to make decisions—no matter where they are, what device they use, or what role they play in the organization.

(Continued)

Advancing to an Industrial Digital Data Infrastructure 7

> **WHAT IS A DIGITAL DATA INFRASTRUCTURE? (Continued)**
>
> Although people often think of data as a value for a given transaction, the digital data infrastructure collects data by the second from sensors. This adds the context of time—*when* different conditions are occurring—and enables framing events (such as operational conditions) by their start and end times. A true digital data infrastructure captures data, time, and events in *real time*.

Bill told Peter that he wanted to achieve this implementation in one business year. "Jeff is on board with this," Bill had told Peter during the interview process. "I have seen this done before. The consultants come in and share what they see about us. The changes we need to make are clear. But it's up to the people here to make them happen."

"Change is hard," Bill continued. "But the knowledge is in the people. They just need to be motivated to adopt a more proactive way of doing things."

To Peter, the consulting firm's recommendations were a guide. A digital data infrastructure simplifies the integration of data from many sources. It enables the collection of real-time data streams from sensors and can classify the operating conditions by time and type of event. A digital data infrastructure also facilitates data analysis to measure the performance of all the processes in a plant. These operational metrics are used for continuous improvement and innovations of production quantity and quality, equipment availability, operating costs, and environmental and safety regulatory compliance.

The consultants' report concluded that by successfully implementing the EIDI program, ProcIndustries management could expect its performance to improve substantially and thus improve the company's competitive position going forward.

Peter understood that it was up to him, building on the consultants' business case, to identify the best way for ProcIndustries to implement an EIDI that would transform the company's most important workflows. To succeed, the digital data infrastructure would need to be integrated with the company's other new information systems that monitor the intake of resources and production processes, including

- Distributed control systems (DCSs), which integrate the process control systems and programmable logic controllers (PLCs) so that operators can supervise all process units in an industrial complex;
- PLCs, which are used for running a sequence of events in process and manufacturing operations;

- Supervisory process control systems, which ensure process units run according to production targets and in an environmentally safe manner;
- Enterprise resource planning (ERP), which tracks all transactions;
- Laboratory information management systems (LIMS), which provide for the product quality monitoring of chemical processes at the plant;
- Tank gauging, logistics, planning, and scheduling; and
- Other systems that provide data that the company accesses from suppliers, customers, and other sources.

It was time for Peter to figure out how to lay the groundwork for a return on the company's investments in the consultants' work. Peter was given two months to devise a plan that would take about two years to implement. Because he had never before worked for an oil company, Peter's focus would be the practical ways to build an EIDI for a large oil refinery enterprise.

Peter thought that he would need to communicate a vision about the importance of the digital data infrastructure that plant managers would accept. He knew that he would need to work with the information technology (IT) professionals at the refinery, people who most likely did not know about real-time data and events in the terms that a processing industry would use. Unlike IT-related events, real-time events in the production process are not always easy to define; it takes special care to identify and track them. Peter expected to have important discussions with the IT department around this issue.

Peter also understood that he would have to work to implement a digital data infrastructure with colleagues across the entire range of functions at ProcIndustries, from operations and maintenance to regulatory compliance.

Peter understood from his briefings that ProcIndustries was like many manufacturers: Process engineers spend a tremendous amount of time figuring out how to translate data into daily performance reports. Peter wanted to use his past experience in energy and manufacturing to show his new colleagues how a digital data infrastructure could help them monitor selected events, and trigger actions within the refinery. These events can then be ranked using management quality initiatives, such as Six Sigma. Data visualizations, such as a Pareto chart, can help colleagues prioritize events to address.

By adding up many actions in one section of the refinery and coordinating them with actions in the other sections in the four-section refinery, the entire organization can increase productivity. The organization can more precisely calculate (in dollars) resources consumed and costs incurred (in gasoline not produced). They can also see energy consumed and raw materials used versus projected levels.

Because the data brings visibility to operations, adjustments are enabled. For example, when an operator has an issue, the operator can use the analysis delivered by the digital data infrastructure to determine what actions to take and engage a maintenance crew or the refinery's operating support team. And because the operations and maintenance teams are working with the same digital data infrastructure in real time, they will pick up the issue and take action. If maintenance staff members have questions, they can request a video chat with an on-duty process engineer who can discuss options for resolving a problem or improving an outcome.

For Peter, a major challenge in this journey is to move ProcIndustries from a reactive to a proactive mode—from treating equipment problems as repairs to helping refinery staff see opportunities to prevent downtime and improve productivity (see Chapter 6).

At this point, however, Peter is focused on more urgent matters:

- Understanding the current state of data management at the South Texas refinery
- Streamlining the current refinery data infrastructure
- Developing a cause-and-effect model for refinery operations that identifies what causes problems (such as actual production levels that don't match planned levels) and how to effect improvements
- Integrating data feeds pertaining to refinery operations with ERP and other systems at ProcIndustries

Peter will learn, share, and build consensus with his team during this journey because, without the collaboration of his new colleagues, little will happen.

The Current Data Infrastructure

Peter spent the next several days listening and charting his understanding of the current business practices at the ProcIndustries refinery where the explosion occurred (see the box "Information Gathering"). He took extensive notes and kept the following questions foremost in his thinking:

- What are the business processes in the oil refinery plant that use data and how is that data used?
- Are there ways to change the work processes to make them more aware, or respond sooner, to problems that occur? Are there opportunities to run things more effectively?
- What are the steps needed to take to make these changes a reality?

INFORMATION GATHERING

Any enterprise that embarks on a digital data infrastructure project will face the challenge of understanding exactly how things work so that data creation, collection, and analysis are relevant to the organization's work and so that everyone involved can agree on what the data shows and how to act on the data analysis. This consensus creates the momentum for positive change.

The next day, the South Texas refinery held a meeting to investigate the current situation at ProcIndustries. The plant manager, Tom Jordan, introduced Peter to the operating team at this facility. For Peter, the meeting revealed some important facts about ProcIndustries' current capabilities and some important gaps in the plant's ability to gather and manage information (see the box "Collaboration Is Key"; Figure 1.1).

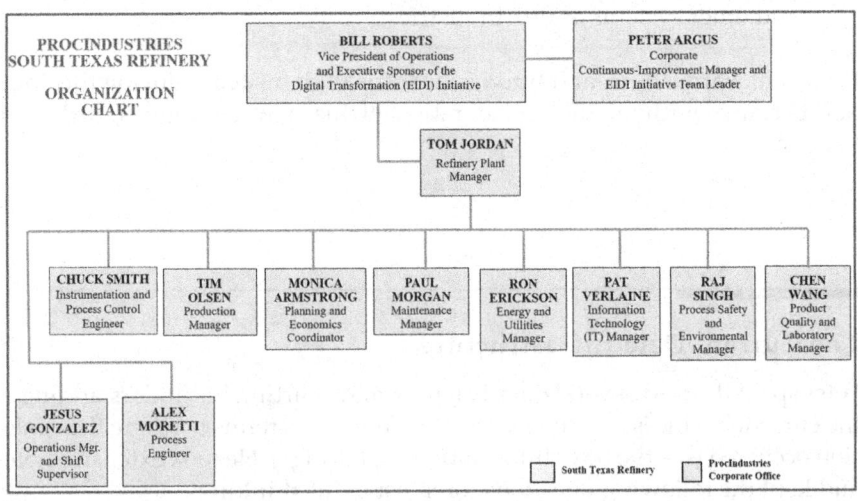

FIGURE 1.1
ProcIndustries organization.

COLLABORATION IS KEY

In every industry, people must collaborate to create an environment in which workers in different roles share data about their work and find ways to improve production. The following table illustrates the roles at the ProcIndustries refinery.

Name	Job Title	Role and Relevance to Digital Data Infrastructure Project
Peter Argus	Continuous-improvement manager	Oversees the digital data infrastructure implementation with the goal of finding ways to have people apply their knowledge and creativity to solve business process problems.
Tom Jordan	Plant manager	On-site facility manager and team leader. Leads on-site staff (operations, production, and maintenance).
Tim Olsen	Production manager	Charged with meeting production plans and schedules. Ensures that the refinery runs in the most profitable and efficient manner given available resources. Aligns the refinery profit objectives with ProcIndustries' corporate strategy.
Paul Morgan	Maintenance manager	Manages maintenance activities and budgets. Processes work orders and schedules their execution. Responsible for providing instrumentation and electrical support to operations.
Monica Armstrong	Planning and economics coordinator	Defines weekly production plans to achieve company goals. Integrates information on raw material inventories, including different types of crude oil, with production process capabilities and constraints.
Chen Wang	Laboratory manager	Analyzes the conditions of production processes in the refinery as well as product quality. Interprets results for other refinery staff members.
Ron Erickson	Energy and utilities coordinator	Manages the consumption of steam, electrical, and water utilities for the refinery.
Raj Singh	Process safety manager	Supports refinery compliance with workplace safety regulations from the Occupational Safety and Health Administration (OSHA). Monitors and records change-management activities for all operations and equipment maintenance activities.

(Continued)

COLLABORATION IS KEY (Continued)

Name	Job Title	Role and Relevance to Digital Data Infrastructure Project
Pat Verlaine	Information technology (IT) manager	Responsible for IT assets, networks, and activity at the refinery. Coordinates with corporate IT.
Alex Moretti	Process engineer	Supports engineering for one section, or process unit, of the four-section refinery. Monitors operations of the unit, including product quality, and informs the business team of results. Troubleshoots operational problems as they arise.
To be hired or promoted from within	Refinery operational intelligence manager	Reports to the plant manager and vice president of refinery operations. Key player in managing employees during the implementation of automation systems. Assigned to implement integrated information systems at the refinery and to identify opportunities for implementing new technologies to increase the effectiveness of the operation as it applies to corporate goals.

Understanding the Barriers to Success

The Process Engineer's View

Tom suggested Peter work with Alex Moretti to understand the role of the process engineer in charge of the production systems. Alex explained that the South Texas refinery has several process control systems. For example, the refinery has a DCS, which uses communications networks to command production during several points of the manufacturing process. Peter found out that the refinery had several process control systems implemented over the years in efforts to enhance the oil and chemical production process.

The main production line was well instrumented. Peter learned that the refinery has a process data historian, which provides real-time streaming data (also known as a *time-series database*) to the process engineers, who use the data to generate their daily spreadsheet reports about production at the refinery. However, each of the process engineers creates their own

Advancing to an Industrial Digital Data Infrastructure

spreadsheets using their own formats. All of these spreadsheets with daily, averaged data are emailed to the administrative office to be consolidated and manually entered into the ProcIndustries ERP system.

But Peter also learned that the refinery's energy supervisory control and data acquisition (SCADA) system was a completely isolated system. This was a surprise to him because he had expected it to be integrated with the refinery's process data historian. In a facility like the South Texas refinery, measuring the specific energy consumption for every unit operation involved in the production process is a key business metric. Having daily or average energy consumption rates is not specific enough to evaluate the efficiency of production processes. Peter knew this lack of visibility would make it impossible to evaluate changes in the production process that related to equipment idling or malfunctioning.

The SCADA system was not the refinery's only data silo (see Table 1.1). Alex explained to Peter that he collected data for the production lines from their DCS local historian, which is not integrated with the refinery's other systems. Once Alex collected this data, he then had to query the LIMS to get the quality data for the day. Alex also gathered more information, using a printed daily report from the utilities SCADA system that showed the refinery's daily energy consumption. He added this information to a spreadsheet he shared with the production team.

TABLE 1.1

Plant Systems: Current Status Versus Peter's Vision for the Future

Data Source	Current Status	Future Goal from Investments in Digital Data Infrastructure System	Business Benefit of Future Goal
Data historian	Provides real-time data for spreadsheet reports. Data is archived offline after two months, making it unavailable for analysis.	Real-time data feeds to asset- and time-context database. Provides context data for production, quality, maintenance, energy, and environmental regulations. Enables data analysis based on operational events and over longer time periods.	Data analysis identifies areas for process improvements, leading to productivity gains.
Distributed control system (DCS)	Separate, disconnected systems provide data about production processes.	Connect these systems to the plant network. Upgrade technology to synchronize all data sources.	Enable plant to employ performance monitoring and instrument validation to confirm equipment is running optimally.

(Continued)

TABLE 1.1 (*Continued*)
Plant Systems: Current Status Versus Peter's Vision for the Future

Data Source	Current Status	Future Goal from Investments in Digital Data Infrastructure System	Business Benefit of Future Goal
Laboratory information management system (LIMS)	LIMS system was recently upgraded; need to collect chemical analysis data from the production process.	LIMS system interface should be upgraded to integrate data analysis into the plant information data system.	Faster report generation. Additional time to correlate process data and quality data for all unit operations. Set stage for future systems with advanced analytical technologies like machine learning.
Supervisory control and data acquisition (SCADA) system	SCADA monitors utilities energy usage. The old system requires upgrading to accommodate additional measurements and is not integrated with the plant data historian.	Integrate with plant data historian and add additional sensors to all motors in the plant.	Provide real-time energy data to the plant data historian for energy management. Set stage for future integration with smart grid.
Production reports based on raw material usage	Time consuming because data does not have context. Performed manually and locally in spreadsheets by one individual.	Reports become part of master plant information asset framework to gather and analyze all data received from process units in the plant.	The reports need to be obtained from one source for analysis and evaluation of conditions. More effective analysis to identify process improvements and diagnose problems. Automate real-time alerts about plant operations.
Energy report on electricity consumption	Reports show usage for whole plant, rather than on a unit or process basis.	Utility consumption is accurately measured for each process unit with specific usage by work shift, product, and operating mode.	Identify trends in consumption, resulting in optimization, and cost savings. Take advantage of unused capabilities in the existing system.
Electronic document management system	Collection of process flow diagrams, piping and instrumentation diagrams, manuals, material data safety sheets.	Information is available via plant information and asset framework system, document-sharing system, and web-based data visualization and analysis tools.	Operate the plant in more efficient and safe manner by providing necessary information to all plant staff members on demand.

Advancing to an Industrial Digital Data Infrastructure 15

It was a tedious task that took Alex several hours per day. He said it was a commonplace that several engineers were days late in adding the production line data into their spreadsheets. The problem worsened when they were sick or took vacation. In addition to collecting data from various sources for reports, Alex explained he had also been dealing with changes to the raw materials coming into the refinery.

Peter asked Alex about process flow diagrams, which refining and other process industry companies use to document production flow and equipment used in the refinery. Alex noted that he kept his process flow control diagrams updated in the Microsoft Visio application together with a spreadsheet of the process diagram containing the process data. Alex explained to Peter that other engineers at the refinery had created similar process flow control diagrams using AutoCAD, the company's official drawing package. Peter asked if this meant Alex and other colleagues shared the same view of the refinery's processes or if they had different views. "I think they vary somewhat," Alex said. (See the box "More about Process Flow Control Diagrams.")

MORE ABOUT PROCESS FLOW CONTROL DIAGRAMS

The process flow control diagrams (PFCDs) are the schematics used for building the plant and their process simulators. These diagrams contain valuable basic information for digitizing a process plant. See Chapter 4 for a discussion on the importance of process flow diagrams to an effective digital data infrastructure.

Listening to Alex, Peter thought that having process flow control diagrams with key metrics for each process unit could be a great opportunity to connect the process units with the new digital data infrastructure tools. ProcIndustries would need a process data model for digitizing the refinery. The diagrams could also be used for collaboration and to identify process improvements.

The Plant Manager's View

Next on Peter's agenda was a conversation with Tom Jordan, the plant manager, to discuss workplace safety documentation, a key element in the refinery's regulatory compliance requirements.

Tom told Peter that the refinery had implemented a partial process safety management system. By this, he meant that the refinery had electronic documentation about the procedures to follow in case of emergencies. However, the process diagrams were scanned and updated only when the process engineers had time to look at the changes.

This meant that Tom had incomplete data and no reliable way to update the data and instructions for emergency situations. Like other information systems at the refinery, staff did the work manually, when time allowed. Although one person may have the knowledge of what to do, it was not necessarily accessible to others when needed. What if refinery equipment, processes used, or raw material ingredients used in production had changed? How could Tom know what to do in an emergency without having a good process flow control diagram or a process flow schematic with the real-time data?

To start digitizing the refinery, Peter explained, they would need to agree on data definitions and a way for communicating about the data at the refinery. Peter insisted that the potential benefits went even further. The process engineers could collect and analyze data in a more granular form. Instead of having a gross average for the day as the only option, they could aggregate the data by operational events as exceptions to routine operations. This would simplify reporting and free engineers to do more analysis related to monitoring processes and equipment and identifying quality improvements.

At the end of these conversations, Peter thanked Tom and the process and quality groups for sharing information and their willingness to collaborate in the redesign of their data infrastructure.

The Planning and Economics Coordinator's View

Back at the refinery the next day, Tom introduced Peter to Monica Armstrong, the planning and economics coordinator. Tom spoke highly of Monica, mentioning that she was a chemical engineer with five years of experience who was studying nights to get her MBA at a local university. Peter was on the hunt for allies, and he felt that Monica's position and her pursuit of more education could make her receptive to helping him when it came time to do the change-management work required to implement a digital data infrastructure project.

- In the meeting room, Monica quickly got down to business, explaining to Peter that the refinery's planning system was based on several factors:

- An economic analysis of raw material inventories
- A production forecast provided by marketing
- Usage reports on power, water, and other consumables (such as steam, water, electricity, fuel, chemical solvents, hydrogen, oxygen, and nitrogen) based on historical averages

Monica was in charge of running their refinery's linear programing (LP) optimization application. The LP system is designed to maximize the refinery's production efficiency. Monica prepared the daily plant production schedule based on the previous day's inventories, process lineup (identifies which process units receive what input materials and where to store the products), product pricing, and the marketing forecast.

Every day, Monica prepared the daily schedule and handed it to Tim Olsen, the production manager. Tim then integrated this report into a set of daily activities to determine the refinery's schedule for the day by

- Reviewing the daily schedule Monica prepared during a morning meeting with the leaders of the operations crew at the refinery and Chen Wang, the laboratory manager in charge of quality;
- Examining the prior day's production activity with information provided by the operations team;
- Checking the refinery equipment maintenance schedules;
- Reviewing inventory reports from frontline workers connecting process units to tanks holding raw materials;
- Modifying Monica's preliminary schedule based on the current refinery situation; and
- Providing a final validated schedule in the form of a spreadsheet to refinery operations staff.

The Production Manager's View

Peter had a follow-up conversation with Tim to learn more about how the refinery teams collected and used the operational data for energy and resource consumption. Peter asked Tim if plant managers had a way to access process information and evaluate production variances. Tim explained that the plant managers use a spreadsheet provided by the process engineers for the four areas of the refinery. However, these reports did not contain detailed total consumption of electrical power used by the four refinery processing units. But they did have good measurements on total refinery consumption of fuel, including gas and diesel.

Peter thought Tim's spreadsheet reports were adequate in terms of yield and quality. However, they lacked valuable information to define the operating costs or minor losses that occurred during fluctuations in production, such as when process units were running during the day, when the units were idle, or when producing slop oil that was not suitable for sale. Peter knew that capturing the data about these minor losses would provide useful information for improvements at the refinery, including better production planning, equipment maintenance and availability, and improved process controls.

During their discussion, Tom, the plant manager, walked in. Tom had been working with the repair crew that was examining the refinery's transformer and assessing the economic consequences of the recent power failure. The crew was able to install a new, bigger transformer and restore electrical power to the sections of the refinery that had been down for the last two days.

Peter and Tim provided a summary of their discussion so far. Process engineers reported that they spent about 50% of their time creating reports that contain average daily production information to assess the yield and daily performance variances. However, between running the reports and handling the process improvement projects, the process engineers did not have time to explore opportunities that could improve the operating costs of the refinery.

"What do you suggest we do to improve the situation?" Tom asked. "You have some experience in this."

Peter told him the refinery had to integrate its energy and laboratory systems into a digital data infrastructure to provide the data to operations, maintenance, and planning and economics. The supply of digital information in real time would enable the refinery's staff to monitor production processes and unusual events. The staff would be able to create reports on specific issues at the desired level of detail using information dashboards.

"But this future state," Peter went on, "requires equipping engineers, managers, and others at the refinery with tools and training for continuous learning. Right now, everyone is using spreadsheets that speak to yesterday or last week. They need to begin using business intelligence tools to visualize the data that reflects current conditions. I believe the team needs to have the tools to become more real-time data driven for quicker operational improvements."

Peter worked to develop a business case to reengineer the refinery's current system (see the box "Building Blocks for Success").

BUILDING BLOCKS FOR SUCCESS

Bill Roberts, the vice president of operations at ProcIndustries, hired Peter to create a data-driven decision environment at the company's South Texas plant. Starting work on such a project takes careful planning—no matter the industry. Here are some essential requirements for a successful digital data infrastructure project:

- Executive sponsorship
- A single project manager who reports to the corporate champion
- Clear communication with everyone who will be impacted by the project
- Explanations of the project's business benefits (and career benefits, where applicable) to those whose work is likely to change
- Training on how to use new tools to improve the likelihood of success
- Examples of best practices with reusable templates for business processes, where possible
- The creation of a virtual or on-location central support team who can assist locally and act as a center of excellence

Peter told Tom the company would have to invest in the time and resources required to map the technology to the production operations at the refinery. Plant engineers and other staff members would need to learn how to build shared, reusable templates for each of the process units in the refinery.

"No more isolated spreadsheets," Peter said. "Information about events at the refinery will be standardized and analyzed by our knowledge workers at the right operational time intervals. This will improve everyone's awareness of operations at the refinery—including power surges like the one that caused the explosion."

He continued, "With the capabilities of the new system, you will be able to see changes in specific electricity consumption and identify the need for pumps, compressors, and other equipment upgrades before they fail."

Peter then explained it was still early and he had more information to gather about the refinery's operations and maintenance. Tom said the next person to hear from was Paul Morgan, the plant's maintenance manager.

The Maintenance Manager's View

Peter was particularly interested in Paul's perspective and stance toward the project. Paul had worked at this refinery for the last 15 years, and he knew every piece of equipment in it. He was involved in the company's move to implement an ERP maintenance system about five years ago, and Paul was aware that ProcIndustries was looking to change the way it managed equipment and processes in the plant through more and better data analysis.

Peter invited Monica to attend this meeting. In the meeting, Paul explained to Peter and Monica that the maintenance crew was always busy repairing equipment. He said that the operations crew was always pushing the units—compressors, feeders, and pumps—close to their operating limits and equipment ratings to meet production schedules.

Paul said his team used the ERP system to define equipment repairs and to generate work orders for these repairs, but the ERP covered only a section of the refinery that was recently refurbished. For that section, maintenance crews had data about the number of hours machines were operating. For the other sections, the maintenance personnel had the option to manually enter process equipment operating conditions, if and when those conditions were available. However, the maintenance staff did not have easy access to real-time data or historical trend data. The result was that they performed equipment maintenance on a time-based schedule predicated on their experience with the equipment.

Paul showed Peter the refinery's downtime reports. These reports spelled out how Paul's maintenance team had to adjust their work schedule so that different refinery sections met daily production requirements.

The maintenance crew worked more like a fire brigade, rushing from one crisis to another, Peter thought. These up-and-down cycles also meant the equipment would deteriorate faster than if the maintenance and operations crews had a steady flow of information, such as how long machines had been running and the real-time health status of those machines. Peter thought this could be an opportunity for Paul to start digitizing equipment and letting the maintenance people configure notifications.

"The data is there. It's collected, but it's not readily available for our group to analyze or to be used by anyone else," Paul said.

Peter said he thought these problems were not unusual. He noted that this lack of visibility into maintenance, combined with the other issues he had heard so far, was building a case to justify the needed investments—the purchase of new sensors, networks, and software upgrades—required for digital transformation.

Peter sensed he would have to find common ground with Paul. Peter needed a way to explain to him that once they installed and classified the major pieces of equipment, the workflow would be easier, given timely access to equipment behavior. Peter would train the team to help them understand the digital data infrastructure project and each person's role as it related to the project going forward: what they could expect, how to utilize it, and how it would benefit them. (See the box "Building a Business Case for Digital Data Infrastructure.")

> **BUILDING A BUSINESS CASE FOR DIGITAL DATA INFRASTRUCTURE**
>
> Collecting all the pros and cons for the project allows a project leader to make an argument that is both specific to the company's needs and relevant to its broader market position. Here are some key questions to answer:
>
> - Why are these investments important?
> - What will be the payoff in terms of costs avoided or revenue gained?
> - What difference does this make in the competitive position of the company?
> - What are the costs and potential benefits of taking no action?
> - What are the sustainable risks to the company of maintaining the status quo?

The IT Department's Role

After Peter, Paul, and Monica finished their conversation, Peter asked Monica to join him in a conversation with Ron Erickson, the energy and utilities manager, and Pat Verlaine, the IT manager. Ron had been with ProcIndustries for 10 years. Ron told Peter that their energy report was built several years ago consolidating the overall energy consumption used by the refinery. The energy report showed the total consumption electricity for the refinery but did not specify usage by individual process unit. Peter noted that the production report did not include process performance analysis for process engineers, machine operators, and managers.

Peter asked Pat about the IT department's role at the refinery. It quickly became clear that while Pat knew much about IT systems at the refinery, he did not have a good understanding of the process. For example, he didn't know the sensor-naming convention used in the aging SCADA system. That meant when someone asked for help to assemble a spreadsheet report about refinery production, Pat was unable to find the data needed to create it. No one had established a data model of the refinery, which mapped the available data to the manufacturing assets.

Ron was frustrated: "I want to know what we should do about this situation. The transformer blew. We're losing significant revenue from lost production time. We need to fix broken compressors. We can fix them, but who knows what will break next? We can guess, but what we really need is a plan."

"We understand the urgency of the situation," Monica interjected. "That's why Peter is here."

Peter shifted the discussion to the digital data infrastructure project: It was time to talk about data mapping for digitizing the refinery. Peter was coming to see himself as the digital transformation coach to turn traditional engineers into digital process and maintenance engineers (see the box "The Engineering Group's Role").

THE ENGINEERING GROUP'S ROLE

In an organization like ProcIndustries, the engineering group plays a critical role not just for making oil-based products but also for the data collection and analysis that will be part of this project. Two roles in particular are essential for the digital plant:

1. Process engineers *define and monitor the operational conditions for production.* They set conditions to meet product specification requirements while also minimizing costs and maintaining environmental and workplace safety compliance. In best practice organizations, process engineers evaluate process improvements and manage needed equipment upgrades.

2. Process control engineers *define and maintain the equipment controls to ensure stable production runs.* They monitor readings from instruments associated with the production process to maintain optimum operating conditions (e.g., flows of petroleum, pressures and temperatures) and ensure that they match target ranges. They monitor process control system performance and add new instrumentation, as needed. They also define when process alerts occur.

"Now," Peter concluded, "we need to adapt the scope of our process and business personnel to bring the operations and IT worlds together. The process of data mapping needs to be done by the process engineers and the process control engineers, and they have to make time to do it."

ProcIndustries has a plant information system but needs to improve it. The information gap caused by data silos affects everyone at the plant. To understand the context of the data describing the operating conditions at the refinery, upgraded and connected user interfaces are needed. There are

Advancing to an Industrial Digital Data Infrastructure 23

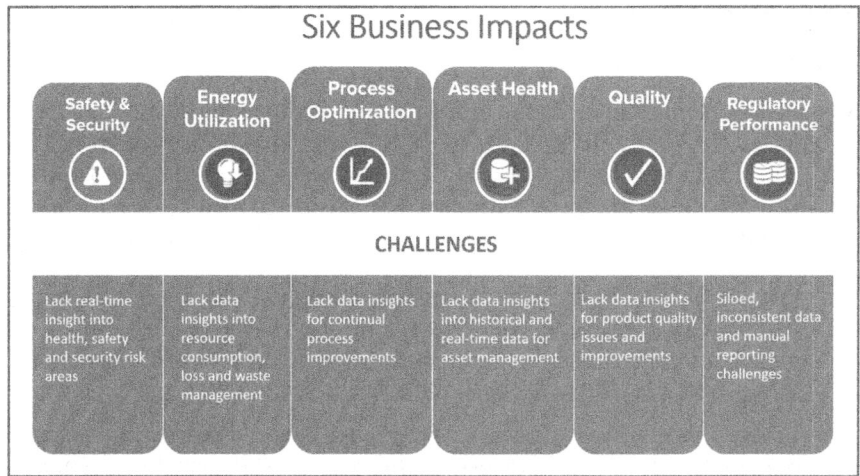

FIGURE 1.2
The major business impacts to be addressed by the EIDI.

essentially six impact areas affected by siloed operational data. Figure 1.2 summarizes the data challenges by operational area at the South Texas refinery.

The Cost of Downtime

Before he left for the day, Peter met with Alex, the process engineer, to find out the operating costs associated with downtime that resulted from pushing units to their limits. They reviewed estimates of how much additional output could be realized if unplanned equipment downtime were reduced.

Together, they determined that an upgrade from the existing plant information system to the proposed EIDI would reduce unplanned downtimes by more than 15% and trim power consumption by 3% to 5%. Peter noted this estimate as another supporting argument for a digital data infrastructure project.

Peter felt encouraged by his talk with Alex, but he worried about Paul, the plant maintenance manager. Peter was concerned that Paul would be reluctant for his maintenance crew to take on the additional work required to support a digital data infrastructure project. It was understandable; the maintenance crews were already overtaxed with equipment problems. The work Peter anticipated meant they would have to connect major equipment to systems that would provide both process and equipment data. That was in addition to their role of maintaining and repairing machines.

Envisioning a Strategy

After a day of meetings, Peter believed that Monica could be an ally for the digital data infrastructure project. She had spoken up for him when Paul got frustrated during the day's discussions. Peter sketched out the vision he planned to share with Monica about the three capabilities an EIDI enables:

1. Proactively detect shifts in process performance. Use statistical tools and process knowledge to identify shifts in performance and provide a structured means of determining the cause of the shifts and, if required, corrective actions to take.
2. Identify process improvement opportunities. Provide sufficient information to identify areas of improvement and new performance indicators.
3. Aggregate data from across the refinery to provide insights into equipment performance. By combining production event data with performance trend information and integrating this data for analytic tools, the team can identify process improvements to be shared with peers and management.

The maintenance crews were not the only ones who needed to perform extra work for this project. Alex, the process engineer, and his team would have to identify the process units and document the data variables associated with them so the digital data infrastructure project could collect the proper data and provide plant staff with refinery performance analysis.

Peter told Monica, "Every business has to work on improvements, but this project we're starting has the potential to be a rebooting strategy for ProcIndustries. EIDI is a technology that can make big changes happen quickly."

Peter identified a series of gaps in the ability of plant engineers and maintenance crews to collect and share data and to comprehend current refinery conditions. Their job functions would change, and with change comes uncertainty and skepticism. He concluded that the team would have to effectively communicate the extensive benefits of EIDI as an "enterprise industrial data infrastructure" so that the teams understand this will greatly benefit them.

Monica agreed. "It's very hard to change our current company culture. Everyone here works hard, but the plant people think that executives at headquarters don't understand their problems." Monica noted that refinery employees might look at Peter as an interloper sent by headquarters to make trouble. "But they are so deep into their own work that it is hard for them to see the bigger picture," she added. "They can't see that there could be a better way of doing things."

Monica pointed out that getting rid of their beloved spreadsheets would be a formidable task. They will be replacing these with a shared view using the EIDI. They will soon be able to create alerts about important refinery events and ultimately use predictive analytics to anticipate issues that need attention (Bascur et al. 2011).

Peter emphasized the importance of digitizing the refinery and said that Monica's modeling expertise would be a great asset in the implementation.

With the EIDI program in place, Monica would be able to fine-tune the refinery's linear programming optimization model using real production data and current refinery capabilities. This would improve the accuracy of the plant production schedules she created, and the effects of using more accurate data would spread throughout the plant. Everyone would have access to operational data on their devices—PCs, laptops, tablets, and mobile phones. Having a digital data infrastructure in place would allow plant engineers to eventually integrate more advanced data analysis tools, such as multivariate analysis and machine learning.

Monica added that there was no doubt if they reduce the refinery production variances, then they would capture top management's attention.

The key, Peter said, was to align planning and scheduling with the daily production results. Having an easy way to aggregate the data from the production, or execution side of the plant, and being able to provide the right information to adjust the LP optimization model—to balance forecasting, execution, and analysis—would enable the plant to execute continuous improvements. When using a common digital data infrastructure as the official system of record, people have a flexible data environment to tackle continuous-improvement initiatives.

"It sounds exciting," Monica said. "Please let me know what I can do to help you make this happen."

Peter left the conversation feeling that he had gained Monica's trust. However, even though the main thrust of his plan for a digital data infrastructure was clear, he knew that his plan would have to incorporate the quality management work that ProcIndustries had already started.

Creating a Breakthrough Vision

When Peter arrived for his meeting with Tom Jordan, the plant manager, and his boss Bill Roberts, there was some tension in the room. Bill, the vice president of operations, was still very concerned about the ramifications of the South Texas plant explosion. He was on the phone trying to coordinate a way to provide the required volume of products with capacity from other ProcIndustries plants.

It had not been an easy few days, Bill told Peter as they sat down in Bill's office. Bill explained that just picking up the phone and asking for more product from the company's three other refineries was a major undertaking. To make things even harder, the other refineries each had their own internal methods to track performance and had very different methods to track daily production.

Before Bill joined ProcIndustries, he worked for a highly regarded large oil company. That company had already implemented a data infrastructure to standardize performance management procedures at several refineries. Bill believed in standardizing and simplifying work processes to enhance collaboration between all teams. He had dealt with issues like complying with OSHA regulations and quality management programs.

The hazard and operability (HAZOP) studies that he had seen in the past would have helped avoid the recent equipment failures at the South Texas refinery. These HAZOP studies had been conducted before newer technologies, such as big data analytics and predictive analysis tools, existed.

Bill believed that the best way to change the culture at ProcIndustries was to use the South Texas plant incident as an opportunity to change. The disruption was a chance to help people understand that the survival of the enterprise—and by extension, the health of their families and the surrounding community—depended on significant changes being made.

"The status quo is over," Bill said. "It doesn't work."

Bill had already supported the implementation of an SAP ERP system across the company. The company was looking into upgrading this system to the SAP in-memory HANA platform for faster, real-time views of operational data. As it stood, the ERP system handled transactions, purchases of materials, traditional maintenance, and accounting. It also provided transactional analysis, but Bill thought it was not enough. The rapid changes in electricity pricing, variances in the costs of raw materials, and complex processes to deal with these changes required additional tools so that people at the operating level could be business improvement managers. ProcIndustries needed a new way for the operating system to tackle these issues. The maintenance functions of the ERP system that tracked operating hours for machines in the plant needed to allow the plant maintenance teams to view real-time conditions.

Assessing Data Infrastructure Maturity

Bill asked Peter what he had learned so far. Right now, Peter said, "ProcIndustries' data maturity is in its infancy." People are accessing data from silos and spending large amounts of time generating reports that do not enable collaborative analysis. It is very difficult to identify process improvements. People cannot even identify variances in production metrics from one day to the next. (See the box "Industry 4.0.")

INDUSTRY 4.0

ProcIndustries' quest to their *digital transformation* is aligned with the principles in Industry 4.0 (Kennedy 2018). Industry 4.0 is commonly referred to as the *fourth industrial revolution*. Industry 4.0 fosters what has been called a "smart factory." Within modular structured smart factories, cyber-physical systems monitor physical processes, create a virtual copy of the physical world, and make decentralized decisions. Over the Internet of things (IoT), cyber-physical systems communicate and cooperate with each other and with humans in real time both internally and across organizational services offered and used by participants of the value chain.

- **Interoperability.** The ability of machines, devices, sensors, and people to connect and communicate with each other via the IoT or the Internet of people (IoP).
- **Information transparency.** The ability of information systems to create a virtual copy of the physical world, also known as a *digital twin*, by enriching digital plant models through the aggregation of raw sensor data into higher value context information.
- **Decentralized decisions**. The ability of cyber-physical systems to make decisions on their own and to perform their tasks as autonomously as possible. Only in the case of exceptions, interferences, or conflicting goals are tasks delegated to a higher level.
- **Technical assistance.** First, the ability of assistance systems to support humans by aggregating and visualizing information comprehensively for making informed decisions and solving urgent problems on short notice. Second, the ability of cyber-physical systems to physically support humans by conducting a range of tasks that are unpleasant, exhausting, or unsafe for their human coworkers.

Peter encapsulated the state of ProcIndustries by defining the aspects of a mature data infrastructure: current status, target state, and maturity status. Peter defined what he called a *maturity state* for data availability, accessibility, correlation, analytics, reporting, and the ability to act on new information received. He summarized the refinery's current status and assigned a maturity status to each area, according to five states:

1. Initial/ad hoc
   - Baseline case
   - Changes are made to solve a problem, but not always based on best practices.
   - The solution has not become a standard operating procedure.
2. Repeatable
   - The solution has become standardized.
   - Everyone is aware of the solution and knows what to do.
     - Usually, this means automated event notifications and workflows when an abnormal situation is detected.
3. Defined
   - Staff and systems work well together.
     - Analytics, event generation systems, system notifications, and workflows combine to automate the identification of abnormal situations.
   - People have access to the data and documentation they need.
4. Managed
   - Continuous training is given, with an emphasis on continuous improvement.
5. Optimized
   - The company monitors the utilization and effectiveness of these procedures and integrates them with process and safety management procedures.

Peter said the goals of the digital data infrastructure project were to address the current deficits and move ProcIndustries to a mature state to reduce the risk of failure, and improve the reliability and integrity of processes and data (Figure 1.3).

As Peter articulated his vision, he suggested assigning a team to learn more about the current activities that were in progress at the South Texas plant and at ProcIndustries in general. It was clear from what he saw so far that everyone had his or her own particular way of doing their job with nonintegrated tools.

Peter observed that the ProcIndustries IT department required some attention. Managers shared a consensus view that while IT had done a superb job at the administrative offices, the company had isolated sectors of the plant level that would require network connectivity.

Tom and Bill studied the information Peter presented. Together they created a table to define current status and target status for the refinery for each of the six areas (Table 1.2).

Advancing to an Industrial Digital Data Infrastructure 29

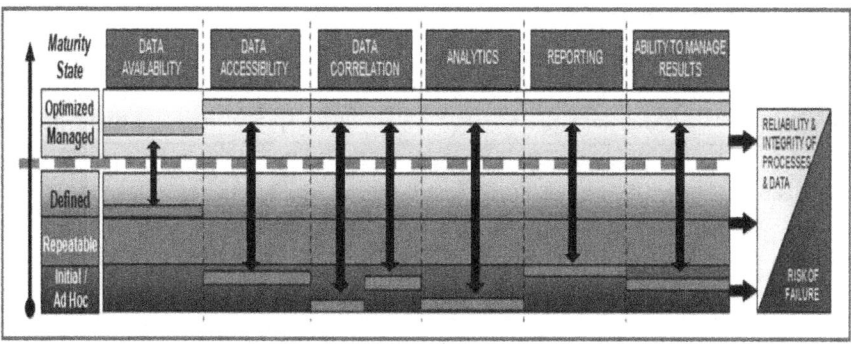

FIGURE 1.3
ProcIndustries data analytics and operational intelligence maturity matrix. Bottom bars are current states; top bars are optimum targets, with the black arrows indicating current deviation from targets.

TABLE 1.2

ProcIndustries Current and Target States

Function	Current Status	Target	Current Status Maturity State
Data availability	Data is available but not aggregated. Data definitions are unique to each plant. No hierarchical data model. Historical data is often lost and difficult to access once offline.	Consolidated data from all available sources is saved in a perpetual historical archive. Data definitions are common and shared across plants.	Defined
Data accessibility	Data is manually collected and often takes excessive time to gather.	Real-time and historical data is easily accessed by anyone in the refinery. Data is also accessible by other systems to automate workflow and business processes. Data is classified according to operating condition or status for proper validation and aggregation, and the operating condition provides a time context for the data (e.g., operating status could be running OK, in trouble, idle, down, in maintenance).	Initial/ad hoc

(Continued)

TABLE 1.2 (*Continued*)

ProcIndustries Current and Target States

Function	Current Status	Target	Current Status Maturity State
Data correlation	Data correlation for modeling requires individual or group heroics. Only "point-in-time" correlations can be made. Correlation across plant sites is nonexistent.	Data is correlated in standardized reports. Subject matter experts can develop and test new data model correlations. Root cause analysis is enhanced to eliminate problems. Cause-and-effect diagrams are available and used together with business intelligence analysis tools. Approved models can be reused and augmented via collaboration.	Initial/ad hoc
Analytics	Analytics require individual heroics. Without validated models, analytics are limited. The capabilities of subject matter experts are not fully utilized.	Analytics are easily performed on real-time data. Subject matter expertise is optimized. Data models capture valuable knowledge.	Initial/ad hoc
Reporting	Reporting is limited to average and totals without aggregation at the desired level of details. Reporting is done manually and covers a point in time. Sharing reports with other plants is not practical or easily achieved.	Real-time and historical trending is available. Contextualized data is aggregated and consumed at desired levels of detail for different roles within the enterprise. Consolidated reporting capability across the enterprise. Data-driven real-time alerts are delivered to staff as needed.	Initial/ad hoc
Ability to manage results	Subject matter experts spend time measuring rather than using the information to achieve results. Departure of personnel creates gaps in knowledge. Ad hoc procedures are implemented to reduce unplanned excursions (departures from the norm).	Subject matter experts are devoted to keeping the refinery in desired operating ranges. Role-based key performance indicators are established and standardized, with real-time values readily accessible. Major excursions are prevented or reduced in duration and impact. Performance is sustainable at all levels. New capital expenditures are justified by data.	Initial/ad hoc

Laying a Foundation

According to Peter, a key element of the digital data infrastructure project is that its impact would grow as plant staff members learn what the system can do. ProcIndustries does not need to implement an EIDI as a single project with a defined end date. The staff will continuously enhance the EIDI by adding new information, new capabilities, and new use cases. It would be a foundation upon which they could build, and the project would begin with a proof-of-concept pilot.

For Bill, this project was an opportunity to reinvent how ProcIndustries planned its production runs, tracked the status of its equipment and asset supplies, and managed the company's compliance with safety and environmental regulations—effectively, ProcIndustries' overall business performance.

Bill shared a diagram that showed three key elements to the group (people, business processes, and technology) surrounded by the enterprise mission and centered on an EIDI (Figure 1.4). He explained that this circle was the ultimate goal and would change over time.

First Step to Action: Convening an Operational Team

Bill suggested that Peter form a new team to support the refinery's change-management program. The team's mandate is to provide leadership, obtain buy-in from the operations and maintenance teams, and help employees learn to use the new EIDI system (see the box "The Importance of the Project Leader").

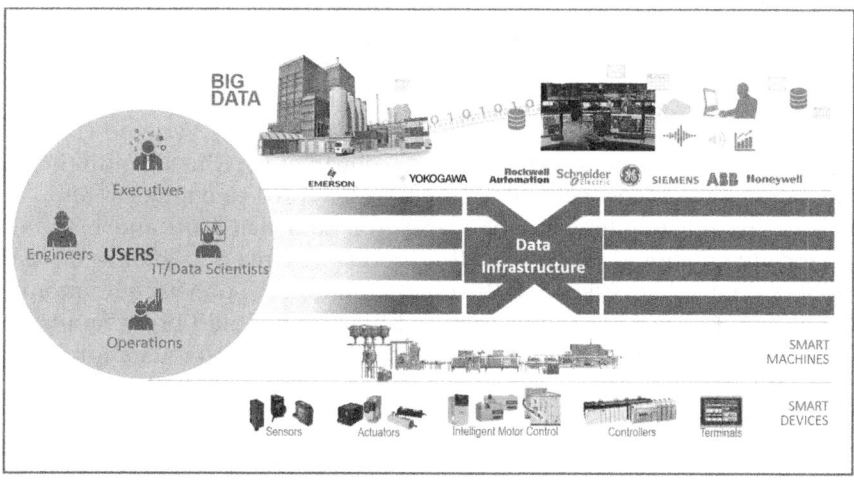

FIGURE 1.4
An industrial data infrastructure connects people, technology, and business processes for continuous improvement and innovation.

> **THE IMPORTANCE OF THE PROJECT LEADER**
>
> Like any project that touches the three core elements of an enterprise—people, business processes, and supporting technology—a digital data infrastructure project requires the establishment of a strong team.
>
> Here is a list of things to consider when assigning a project leader:
>
> - Good listening and communication skills (someone eager to share knowledge and learn from others)
> - Strong coaching skills
> - Curiosity
> - Creative problem-solving skills
> - Business acumen (an understanding of all aspects of the business, not just production gains and losses, but also workplace safety and environmental regulations)
> - Well versed in company culture with a track record of successful projects

"You need people on the team who bring different points of view to the table," Bill said. Bill had worked with a major oil company that had failed to appoint a multidisciplinary team to build support for its data infrastructure program. He wanted to approach it differently this time and thought new technologies would give ProcIndustries the tools to gather and analyze data and present it in a visual format for staff to use.

Bill and Peter settled on the following operational team for the digital transformation: Peter, Tim (production manager), Pat Verlaine (IT manager), Monica (planning and economics coordinator), Chuck Smith (instrumentation and process control engineer), Alex (process engineer), and Paul Morgan (plant maintenance manager). Bill said that Paul and his maintenance team should work with the process engineers to define the data model (what data everyone in the plant would need so they could monitor plant operations). Meanwhile, Tim and his group would bring familiarity with the manufacturing units and could advise the digital data infrastructure project on machine control systems (see the box "The Importance of the Operational Team").

"If this team is successful, it can change the way we run the company," Bill said. "I want this to be a pilot project. When it works, it will show the other refineries how new thinking can solve big problems."

THE IMPORTANCE OF THE OPERATIONAL TEAM

A digital data infrastructure project needs a cross-functional team to lead it. At ProcIndustries, the operational team—with members from maintenance, operations, engineering, and planning—has the job of continuously improving refinery operations to meet corporate objectives. Creating an organizational culture of continuous improvement in a digitized refinery includes reengineering current practices and systems and investing in additional technologies to bring about new levels of productivity.

Engaging Corporate Leadership and Plant Employees

As the meeting was ending, Bill added that it was important that the project team keep the company's top management well informed: "The C-suite is very interested in our overall schedule, early status reports, and tangible benefits."

The digital data infrastructure team would need to conduct a training course on the project and the new operational excellence strategy it represented (Figure 1.5). Bill suggested the team teach employees how their previous training in Six Sigma methodologies applied to this digital data infrastructure project.

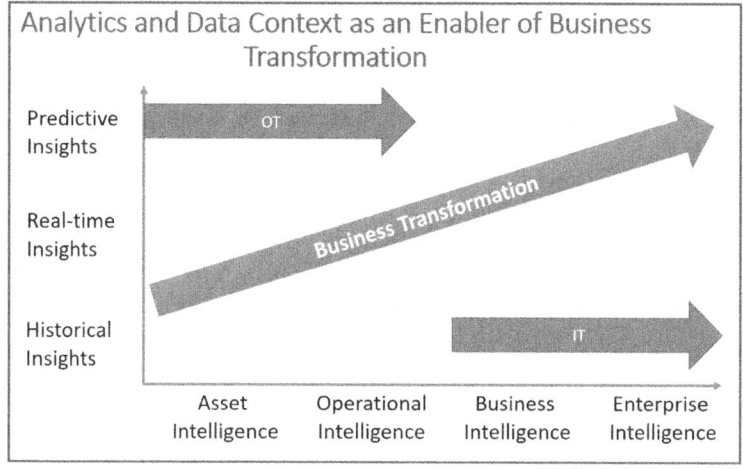

FIGURE 1.5
The OT and IT integration trajectory toward a business digital transformation.

The EIDI project itself would have three stages:

1. **Streamline data access and delivery of data and events.** Roll out a streamlined version of the EIDI to simplify employees' access to information, set up an environment in which they learn about operations, and continuously improve their ability to detect problems.
2. **Integrate data silos for predictive real-time data analytics.** Establish a plant information system framework for consistent representation of assets and equipment in the plant. Determine how to simplify the delivery of actionable information for every asset in the plant. This framework sets the stage for collecting and analyzing data from plant machinery to measure how processes are performing. The employees build real-time cause-and-effect models for predictive monitoring.
3. **Create transformation by integrating operational and business information with time context.** Simplify the refinery's data structure to enable integration with enterprise resource management systems. This will equip business intelligence systems with time-based context. Integrating refinery and corporate business information sets the stage for operational excellence across the company.

What You Should Take Away

Peter Argus, hired as a new continuous-improvement manager, joined ProcIndustries just as the company's South Texas refinery experienced an electrical overload that led to a transformer explosion. Thankfully, no one was injured. However, in the aftermath, Peter and his colleagues discovered a great deal about how the plant runs:

- There is a lack of collaboration across different departments. Each employee has a personal way of managing data that is inadequate for today's business conditions.
- Disparate systems, including data sources contained in legacy systems, prevent plant management from clearly viewing integrated processes, equipment performance, and lab quality data, making problem solving difficult and time consuming.
- The lack of data makes it difficult to align refinery production plans and the execution of those plans. Raw material and utility costs often vary, and existing planning processes have difficulty accounting for these variances.

- There is no real-time monitoring of refinery critical equipment performance, resulting in unscheduled downtimes and excess operating costs.
- Staff at the plant needs training to develop the requisite skills to use a digital data infrastructure.

Bill Roberts, the vice president of operations who hired Peter, assigned him and a team of managers from the plant to prepare a proposal to implement a digital data infrastructure. ProcIndustries management approved an initial pilot to roll out a streamlined version of the digital data infrastructure project, stating it will enable the company to comply with workplace safety and environmental regulations. It will also allow personnel to analyze refinery-wide data and share that knowledge across all areas of the refinery. If the EIDI initiative is successful at the South Texas refinery, the company plans to scale it across the enterprise.

References

Bascur, O.A., Hertler, C., and Wong, G. 2011. "Improving sustainability strategies in industrial complexes: Integration and collaboration." In *Proceedings of the EMC 2011*, ed. J. Harre. Lower Saxony, Germany: Clausthal-Zellerfeld.

De Geus, A. 1988. "Planning as learning." *Harvard Business Review* (March/April). hbr.org.

Kennedy, J.P. 2018. "The driving force of industry." *Scale and Scope* (September/October): 34–37. www.isa.org/intech/20181005.

Kennedy, J.P. 2002. "Just an IDEA." Internal communication, OSIsoft.

Additional Reading

Bascur, O.A. 1988. "A control data framework with distributed intelligence." *Advances in Instrumentation* 88:1553–1169.

Bascur, O.A., and O'Rourke, J. 2019. "Measuring, managing and transforming data for operational insights." In *Smart Manufacturing: Concepts and Methods*, ed. M. Soroush, M. Baldea, and T. Edgar. New York: Elsevier, forthcoming.

Bascur, O.A., and Soudek, A. 2019. "Grinding and flotation optimization using operational intelligence." *Mining, Metallurgy & Exploration* 36(1):139–149.

Botin, J. 2009. "Integrating sustainability into the organization." In *Sustainable Management of Mining Operations*, ed. J.A. Botin. Englewood, CO: Society for Mining, Metallurgy & Exploration. pp. 71–132.

Gonzalez, F. 2014. *Reinventing the Enterprise in the Digital Era*. Madrid, Spain: BBVA.

Kennedy, J.P. 2019. "Retrospective: Then and now—looking toward the future." *Hydrocarbon Processing* (January). www.hydrocarbonprocessing.com/magazine/2019/january-2019/columns/retrospective-then-and-now-looking-toward-the-future.

Kennedy, J.P., and Bascur, O.A. 2002. "Influence of computer and information technology on process operations and business processes—a case study." In *Chemical Process Control VI: Assessment and New Directions for Research: Proceedings of the Sixth International Conference on Chemical Process Control*, ed. J.B. Rawlings, B.A. Ogunnaike, and J.W. Eaton. CACHE Series, Vol. 98. New York: American Institute of Chemical Engineers (AICHE). pp. 7–11.

Lochner, P., et al. 2006. "Integrating sustainability into strategy (ISIS): A process to inform sustainability strategies, framework and reports." *Presented at the Annual National Conference of the International Association for Impact Assessment*, Johannesburg, RSA.

Maister, D. 1977. *True Professionalism*. New York: The Free Press.

Senge, P.M., Kleiner, A., Roberts, C., Ross, R.B., and Smith, B.J. 1994. *The Fifth Discipline Fieldbook: Strategies and Tools for Building a Learning Organization*. New York: Doubleday.

2

Building the Foundation

Continuous measures are one those factors that can be measured on an infinitely divisible scale; weight, height, time, temperature, pressure, ohms, money, energy, flow. *You can't manage what you don't measure.*

Peter F. Drucker

Chapter Overview

In this chapter, you will learn how the South Texas refinery digital transformation team plans for implementation of the enterprise industrial data infrastructure (EIDI). Rather than focusing on the technology, the team reviews current workflows, how data is stored and consumed by refinery personnel, and how refinery operations can be continuously improved by addressing all functional areas, including safety and environmental issues, management reporting, and increasing visibility to all workers. The team will unveil their plans for how various refinery and corporate teams will access and share the information to improve operations and reduce costs. This chapter also introduces how other ProcIndustries refineries and corporate support teams will ultimately benefit from an EIDI implementation.

The View from the Refinery Manager's Office

The last month had been interesting at ProcIndustries. The South Texas oil refinery, which Tom Jordan managed, suffered a major disruption that prompted the company to reexamine just about everything it does and how it does it. Tom was frustrated and upset about the refinery's inconsistent operations, but he kept thinking about the determination shown by Bill Roberts, the vice president of operations. Bill really wanted to improve plant operations and was willing to finance that initiative (see the box "An Operational Excellence Program"). Now Tom was on his way to a meeting called by Peter Argus, the new continuous-improvement manager at ProcIndustries. Peter wanted to meet with the newly formed digital transformation team to begin their work: transforming the refinery culture to a data-driven practice (Duggan 2012; Miller 2014; Gonzalez 2014; Rogers 2016).

AN OPERATIONAL EXCELLENCE PROGRAM

An operational excellence program requires an agile, adaptive organization that helps people to be active, learning participants in the organization and, as a result, creates a culture of learning and continuous improvement. Specific steps to develop an operational excellence program include the following:

1. Define common long-term goals for the enterprise.
2. Articulate how systems and business processes enable people to simplify their jobs.
3. Provide access to data to utilize data relevant to workers' individual roles and responsibilities.
4. Ensure that people use a common set of definitions for the data they are interpreting.

After an earlier series of discussions with Peter, Tom realized that the oil refinery had data locked in legacy applications that people were not using. To unlock that data and add new capabilities, the company must take proactive steps:

- Install sensors on equipment to fill in missing data gaps.
- Upgrade systems to collect, process, and analyze the data.
- Implement new user interfaces and train people to use them.

These steps, when completed, would create a flow of data from equipment sensors in the refinery to the people making decisions about how best to run the refinery.

To date, ProcIndustries had been averaging data from these isolated systems and storing it in relational databases for after-the-fact reports, which did not allow them to drill down and find root causes of problems. The operations department lacked real-time awareness of whether the refinery was running at optimal conditions or encountering problems.

Operations and production data was captured, averaged, and stored in software applications tailored for a specific function. Operations, engineering, maintenance, and the business planning department had their own data, which may not have matched what others had.

Tom felt the culture at ProcIndustries represented another challenge. He knew employees did not believe the top executives had their best interests in mind. The perception was that senior management was out of touch with the realities of daily refinery work. Top management received reports based on accounting data, such as quantities produced and sold, but did not receive operational reports, such as daily operating costs and variances in planned and actual production levels.

Employees also believed that senior management did not want to spend money to remedy data systems. This perception appeared to be changing with Bill Roberts' support. Top executives did not appear to appreciate the additional work refinery staff had faced to comply with new regulatory requirements for environmental rules, occupational safety and health, and public utilities. If implementing a new digital data infrastructure system and applications meant asking staff to do more work, Tom was worried that he would have a management challenge on his hands.

Tom also recognized that the cultural challenges for this project went beyond how the refinery departments viewed management. Tom knew the refinery staff members were accustomed to their own way of doing things. Business processes that once worked well were now so entrenched that few could see that they had become obsolete.

To upgrade their current information systems and implement a modern system was a tall and complex order. Tom asked himself, what factors reduce complexity? He listed these factors as he prepared for the afternoon digital transformation team meeting:

- Universal standards that define business conditions and the data used to describe these conditions.
- A common platform for sharing real-time operations information to all personnel, empowering them to make quicker decisions.
- The ability to quickly analyze data and improve how the refinery is operated.
- A flexible information infrastructure allowing ProcIndustries to manage future business changes.

Tom had been investigating operational intelligence and was planning to share some thoughts with Monica and Peter. His view was that operational intelligence would provide the ability to act on insights through manual or automated actions. Peter was creating a leadership vision for their digital transformation initiative.

Identifying Workflow Information Gaps

In preparation for the digital transformation team meeting that afternoon, Bill Roberts, the vice president of operations, asked Tom to present an overview of various functions at the refinery and how they interconnected. Peter, who is leading the digital data infrastructure project, identified areas where they were missing information that could improve operations. Bill proposed that the team should identify missing links and how they affect refinery business performance. By filling in these gaps, the team would recognize what constitutes a profitable refinery. Peter and Monica could use Six Sigma methodologies in a continuous-process improvement program enabled by the enterprise industrial data infrastructure (EIDI). (See the box "Six Sigma.")

SIX SIGMA

Six Sigma, developed in the 1980s at Motorola, is a statistical measure of variation from a desired result. Adapted in many industries, Six Sigma methodology is used worldwide for improving enterprise process management, with the desired end result being "zero defect" commercial processes. The actions drive process improvements, quality improvements, reduced costs, financial gains, and enterprise competitiveness (NIST 2017).

At ProcIndustries, efforts to incorporate Six Sigma methodologies had fallen short. The proposed digital data infrastructure project would address the following:

- Operational efficiency, optimizing energy and water consumption.
- Reducing refinery equipment failures and unscheduled downtimes.
- Improving returns using Six Sigma methodology, which enables companies to identify root causes of waste, thereby improving efficiency and yield.

It was up to Peter to determine the best way to adopt an EIDI, which would lay the foundation for the company moving forward. Very early on in his career, Peter noticed that many companies discard a valuable corporate asset—production history. Since he began working as an engineer in the process industries, he believed that *people who effectively use data can transform their world*.

Building the Foundation

At the refinery, an EIDI would enable

- The standardization of process methods and how people work,
- Sharing of information among different departments, and
- Development of metrics based on real-time data analysis.

Improving Scheduling, Operations, and Maintenance

Bill mentioned that while working at his last job at a large oil company, he explored implementing a continuous-improvement program in order to comply with government-mandated process safety management regulations for handling hazardous chemicals.

"What became clear to me," Bill said, "is that an organization committed to ongoing quality and safety must examine three levels: processes, assets, and people. When I say three levels, I am not saying that they are all equal. The people part is the most important. A successful implementation depends on people, not systems."

"In the traditional vertical organization, each department has to meet goals set independently of other departments," Bill said. "The independence of these functions creates instability. The push to meet departmental goals can lead a team to push equipment to operating limits. They are then likely to suffer unscheduled equipment failures and production down times."

Bill concluded that, in a traditional organization, each department managed its own data to create reports for its own specific functions. Companies organized this way lost a tremendous opportunity to reuse data for multiple, interdependent functions (Rummler and Brache 2013; Harrington 2012). Tom stressed that essential processes spanned multiple departments, but process data stored in disparate systems had to be aggregated.

Tom showed the group a diagram of the major operational process workflows at the refinery (Figure 2.1). Many times, organizations implement business process workflows, not realizing what it takes to implement them at basic operational levels. The diagram shows four major functions and the flow of information between them:

1. The operation and maintenance integrated team manages the feed stock inventories, processes, process safety management, and performance-monitoring information access. They also provide just-in-time training.

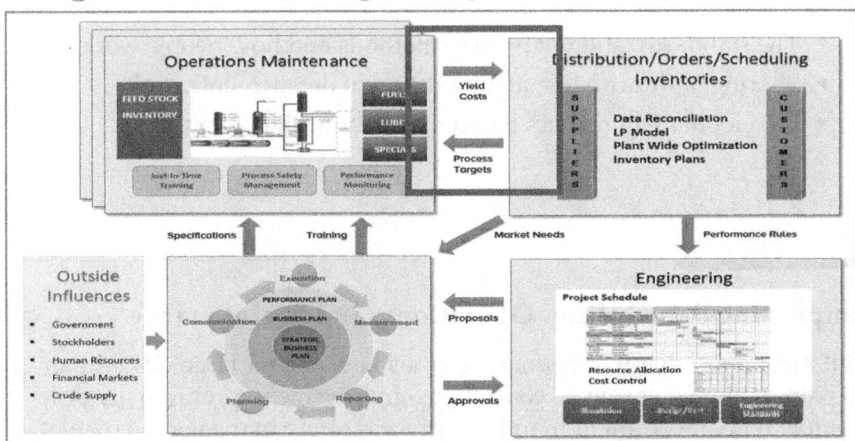

FIGURE 2.1
Business processes workflows diagram indicating exchange of information between various functions.

2. The planning and economics team manages supplier and customer orders using data reconciliation, linear programming (LP) production schedules, plant-wide optimization techniques, and inventory management.
3. The engineering team implements process improvements employing performance rules received from the planning and scheduling team.
4. The plant management team develops the refinery's strategic business plans.

The group discussed how they could reduce information gaps and improve business workflows. They examined new ways of using information and collaborating among the various groups. As an example, Tom described the operations and maintenance workgroups. Although team members shared some information such as spreadsheets, and verbally communicated their needs, operational data and equipment data often resided in different spreadsheets. "The data they gather is separate from planning and economics or management," Tom said. "It may work for them, but not for the company overall. Plus, the data is obsolete when they receive it."

Monica drew a square in Tom's diagram to indicate the information gap between refinery scheduling and operations (Figure 2.1). Analyzing operational data and events helps the refinery staff fine-tune the planning model and avoid process and quality constraints. If the operations and maintenance departments share and standardize their operational data, both groups could reduce unscheduled equipment downtimes.

The unifying principle is the data. What is needed is to identify the hidden production and consumables losses by capturing the unit operating modes and automatically aggregating the data into business information. Tom said, "Optimal production targets can be identified based on the quality and availability of our raw materials, our customers' needs, and the true capabilities of the plant."

Bill responded, "Tom, this is a good start, but I can see even more potential here. If engineers spent more time improving processes and less time gathering operational data, they could optimize production-planning models." Bill went on, excited about management seeing a live key performance indicator (KPI) dashboard of current refinery operations: "That would give management visibility into refinery operations, assisting them in evaluating requests from various departments and taking appropriate actions."

Peter showed the group a diagram (Figure 2.2) illustrating how an EIDI enables multiple parallel use by continuously reusing the same data. This would solve myriad challenges in the operational pillars at ProcIndustries (Kennedy and Bascur 2002).

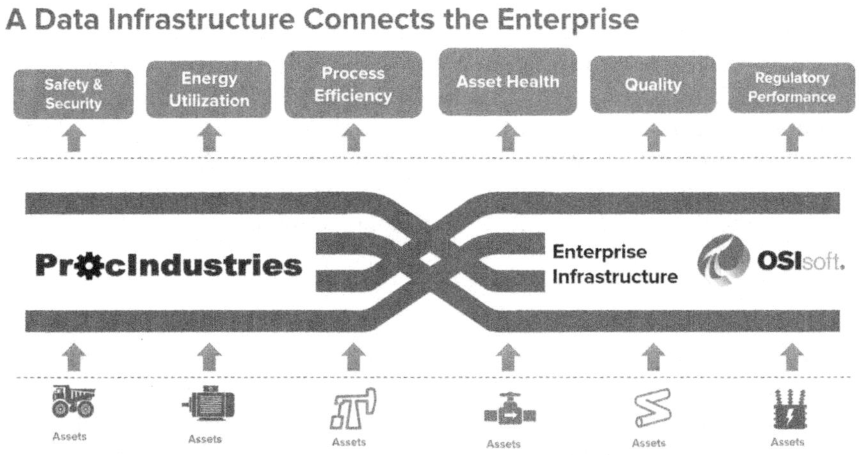

FIGURE 2.2
Real-time EIDI data infrastructure for operational intelligence.

Seeing how the EIDI could tap the enormous talent in his organization, Bill proposed debriefing the entire organization and describing Tom's operational workflow vision. Bill agreed to update the ProcIndustries leadership team on the progress of the company's digital transformation process.

Process Unit Template for Smart Thinking

In preparing for this conversation, Peter and Monica wanted to communicate the importance of sharing a common nomenclature to describe production processes at the refinery. This was a critical issue: Currently, every unit had a different variable name, making it impossible to benchmark production performance and uncover operational constraints.

Peter noted, "The digital transformation team needs to develop a process unit template for the digital data infrastructure project." This modular, reusable template will define the data hierarchy and describe asset content for applications using the digital data infrastructure (see the box "About Process Unit Templates"). It will map the refinery data collected to applications that analyze operations and provide visibility into key areas, such as safety and security, energy utilization, process optimization, asset health management, quality improvement, and regulatory compliance.

ABOUT PROCESS UNIT TEMPLATES

A process unit template is an asset object model that contains the necessary contextual information about an asset as recorded by the digital data infrastructure. This information can be in the context of an important process event or the operational mode of an asset. The digital data infrastructure organizes the data and events for real-time and post hoc analysis. It enables raw data to be transformed into valuable information with the right level of detail. This is discussed in more detail in Chapter 4.

Time is a key variable for the digital data infrastructure project, used for key data aggregation, KPIs, KPI management, predictive analytics, and advanced data analysis. The digital data infrastructure enables the automated recording and analyzing of events, including event start and end times. It then transforms data within these events into actionable information. The key is to use the digital data infrastructure system to automate the capturing and analysis of events. Most relational databases do not handle time-series data well, so ProcIndustries plans to use the EIDI

Building the Foundation 45

with their production data model for machine learning (ML) and artificial intelligence (AI) (see the box "Challenges in Time-Series Data in Refinery Information Systems").

> **CHALLENGES IN TIME-SERIES DATA IN REFINERY INFORMATION SYSTEMS**
>
> Despite the universal recognition that data is critical to smart operations, rarely does operational data inform decisions at all levels of the enterprise. Why?
>
> - Operational sensors and systems produce massive volumes of data and often have process control automation systems that do not communicate with one another. Enterprise data records are often incomplete, fragmented, and frequently inaccessible to many users.
> - As connected assets and increased connectivity lower the barriers to capturing even more data, most systems cannot scale to handle increased data volume.
> - Traditional data archives lack contextual information that add value to data shared throughout the enterprise. Without context, valuable operational information often remains underutilized, sequestered, or unavailable to users unfamiliar with control-system naming conventions.
> - Data collected and stored by isolated point solutions have disparate sources or formats. Reporting, calculations, and roll-ups require manual data entry, are prone to error, and are time and labor-intensive.
> - Many interfaces that collect the data are at the asset location. Accessing information from remote sites or from centralized centers can be challenging, and delay or prevent timely use of valuable information.
> - Connecting data from siloed point solutions, applications and data historians require skilled resources and customized solutions, which adds information technology (IT) complexity and cost.

With this technology, ProcIndustries can isolate refinery events when there is a problem in one of the process units. For example, if a process unit experiences trouble for 15 minutes, workers can mark the start and end times of this event. They can also calculate the total amount of time the process unit is in trouble and estimate how that trouble affected the totals for the consumable variables that were configured in the process unit template.

Every shift can estimate the total amount of consumables (e.g., energy, water, and catalysts) used while deviating from target. They also can identify and totalize the times that the unit is down, idle, or under maintenance, and time on target. With the ability to track data for all of the variables and defined events of interest, they can calculate the monetary value of more quickly resolving problematic events.

Operational Excellence Methods and Tools

The discussion continued. Bill emphasized that the purpose of the company's operational excellence program was to make ProcIndustries more agile in its operations and more profitable as a business to ensure its long-term survival in a competitive marketplace. "The key is to simplify our communications about the basic metrics everyone uses," Bill said. "The digital data infrastructure system will be the refinery's window into these basic metrics."

"To embed knowledge in a modern digital data infrastructure system, we need to use templates," said Peter. Templates set parameters for process analysis in the refinery. And the insights can be applied to improve process performance locally at the refinery level and strategically at the enterprise level.

Continuous improvement, they all agreed, required a structured approach with seven key elements of real-time performance monitoring and control systems:

1. Agreement on critical metrics for production monitoring, product quality, assets, process inputs, and environmental and safety conditions
2. A method for measuring performance
3. Corrective action planning
4. Clarification of roles and responsibilities
5. Regular process reports and review meetings
6. Recognition of progress and achievement
7. Collaboration among different roles at the refinery, including production planning, process, engineering/maintenance, and management

It is particularly important for the company to define measurement methods and continuously monitor them (Bascur and Kennedy 1995).

Building the Foundation

Workflow Management

Peter next explained the importance of workflow management in continuous-improvement programs. "We can improve company performance by combining workflow management with the analytic capabilities of the digital data infrastructure system," he said. "Once we set up the system to collect data and analyze performance, the workflow will be initiated by the digital data system."

Currently, ProcIndustries can track transactions and dollars spent in the refinery, but their existing systems were unable to analyze operating problems that needed to be addressed. Although some of his colleagues were using an enterprise resource planning (ERP) system, Peter wanted to educate them about the benefits of the digital data infrastructure system. "The function of a real-time digital data infrastructure is to support continuous improvement," Peter said. "This is quite different than the ERP system. With the EIDI, we can identify, analyze, and correct operating problems."

Peter showed an illustration that he said was an adaptation of the traditional **D**efine, **M**easure, **A**nalyze, **I**mprove, and **C**ontrol (DMAIC) loop for real-time process improvement (Figure 2.3): "To achieve true change, we need two loops. One is tactical and one is strategic. Metaphorically speaking, the two loops are how the human brain works. The left side does all the

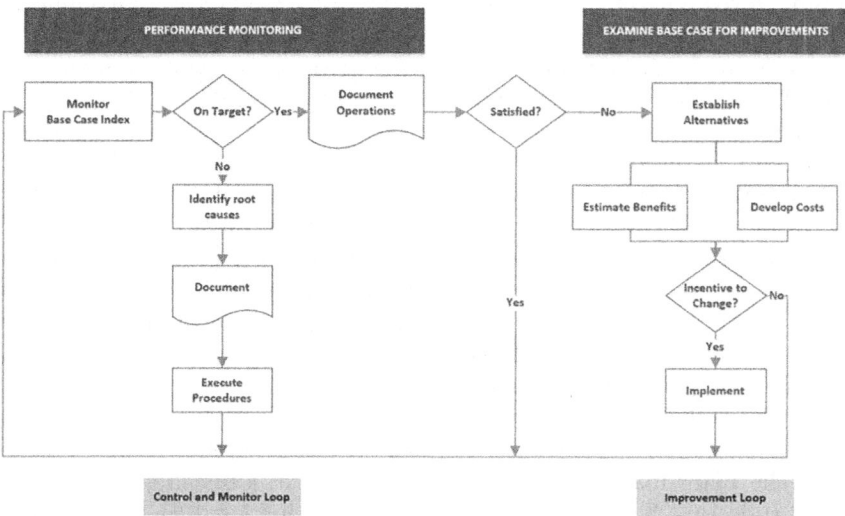

FIGURE 2.3
Real-time continuous improvement and innovation loops.

hunting and linear things. The right side integrates our values to plan and look for ways to get out of trouble if a problem arises."

The figure also shows the post-analysis and improvement workflow activities by the support teams. Together, these series of actions reduce the time required to convert data into answers that support continuous improvement for the organization. Operational continuous improvement is based on two separate loops:

1. Are we on target? (the system monitoring loop)
2. Are we satisfied? (the continuous-improvement and innovations [CII] loop)

The First Loop: Are We on Target?

For a typical plant, the first loop (Figure 2.3 left side) compares the plant unit targets with the current unit metrics daily schedule and uses real-time analytics to generate events that mark the start and end time of operational events. These events are defined by metrics that show activity outside statistical control limits. The first loop is a comparison between the current metrics and target levels expressed at the desired degree of detail. This loop can be performed by the EIDI to catch variances outside the normal targets.

Peter went on: "A fundamental differentiator of a real-time digital data infrastructure or EIDI is that data can constantly be reused as a time-derived variable. We do not use a static variable or a snapshot variable for analysis. A time-derived variable is a critical concept in the way the digital data infrastructure supports our analysis because the time-derived variable enables us to determine conditions in the production process at a specific moment in time. We observe when these events begin and when they end."

Imagine a pump breaking down. The sensors can detect an abnormal activity in the production process and trigger an event through the digital data infrastructure system. Once the digital data infrastructure system triggers an event, the system generates a workflow for the team to resolve the issue and to assign a root cause. Therefore, the digital data infrastructure system is connecting the people in charge of a resolution to the metrics of an unusual event. This triggered event can also send a notification via email, text, or phone call to those in the workflow. This triggered event can be escalated to other teams if so desired.

Peter cited three potential examples:

1. The plant manager, Tom Jordan, receives a notification when the crude unit has been operating below production targets for more than 30 minutes.

Building the Foundation

2. The process engineer, Alex Moretti, is notified if the crude unit starts moving into "trouble" mode.
3. Chen Wang, the laboratory manager, receives a notification if there is a large variation in the production process.

The Second Loop: Are We Satisfied?

Peter said the second loop (Figure 2.3 right side)—Are we satisfied?—is about process improvement and innovation. This is a *people-centric* loop.

In this part of the process, Peter explained, people use their time differently than ProcIndustries staff members have done in the past. Instead of fighting crises or doing busy work like gathering data, they are defining better performance metrics using process models. The loop provides the opportunity to improve a process, to solve a current problem. It is here that people need to question the status quo. Is there something else that we could do to improve outcomes? Ultimately, Peter explained, work on the "Are we satisfied?" loop helps improve the bottom line.

Peter said the "Are we satisfied?" loop is handled by support teams (engineering, maintenance, production, quality, environmental, and safety). These teams act as guides to people in the trenches. They are the process, quality, or maintenance engineers who continuously evaluate long-term performance trends of the operations, instead of receiving incomplete and outdated reports, as in the past. These teams are in charge of recommending suggestions for changes based on their insights. With the EIDI, teams can analyze data at the desired degree of detail and see problems and opportunities that had previously not been able to be detected by the operating teams. For example, the EIDI makes it possible to calculate electricity consumption during an equipment problem. This enables workers to analyze the problem in a finer degree of detail than in the past, and to quantify benefits (such as reduced electricity costs) for improving the situation.

Peter explained that the two loops provide a strategy for process improvement (Table 2.1). The first loop focuses on the day-to-day operations and checks whether they are on target. The second loop seeks improvement strategies based on the belief that performance can always get better. Together, these two loops embody continuous-improvement methodology:

1. Provide operators with consistent quality data and performance indices by checking mass balances, energy balances, and correct flow rates for temperature, pressure, and in some circumstances, chemical composition or specific gravity.
2. Display key metrics for refinery personnel that raise their awareness of overall refinery performance (production, energy, asset, safety, and environmental conditions). These displays, on monitors in areas where people gather, can augment traditional daily reports.

3. Continuously evaluate overall refinery performance that impacts profitability.
4. Quantify a refinery-wide penalty cost when metrics fall below performance targets.
5. Monitor performance of all critical equipment to improve maintenance planning and reduce equipment disruptions, thus enhancing overall refinery availability.

TABLE 2.1

Target and Improvement Loops: Analytic Tools and Bottom-Line Benefits

---

1. Target or control monitoring loop: Are we on target?
   - Provide operations with cause-and-effect diagrams.
   - Provide visibility into operating processes through email notifications, online video conferencing with engineer-mentors, and other enablers.
   - Organize data and documents to find root causes.
   - Enable teams to interpret results using tools such as fishbone diagrams.
   - Review process, production, and quality variances.
   - Provide energy, water, and utilities connected to local smart grid activities.
   - Provide safety and environmental notifications.
   - Improve operator morale through process ownership.
   - Use Pareto analysis to evaluate the most important root causes to eliminate.
   - Use process matrix analysis to identify opportunities for improvement.
2. Improvement and innovation loop: Are we satisfied with the current performance? Review variances and collaborate with planning and economics personnel to modify targets and business rules, where applicable.
   - Identify unnecessary energy costs and implement utility process improvements.
   - Identify operational losses and profits.
   - Identify conditioned-based equipment monitoring strategies.
   - Evaluate mechanical integrity and economic risks.
   - Use business intelligence tools to visualize data as actionable information.
   - Improve employee satisfaction by empowering them to create new solutions with the right data and visualization tools.
   - Drastically reduce the time needed to resolve problems or identify business opportunities.
   - Define short- and long-term process improvement plans and provide proposals to management.

Presenting the Digital Data Infrastructure Project to the Entire Company

"I like what I see here," Bill said. He noted how the two feedback loops would enable the refinery to make important improvement gains. In addition, using the systems to analyze data and events at the refinery and to monitor the flow of production processes represented a chance to bring greater visibility to the operations and production teams.

Building the Foundation

Bill said they would have the opportunity to test this new approach at ProcIndustries within the next couple of weeks. He asked the digital transformation team to repeat this presentation to the entire company using the corporate video conferencing system.

Bill noted that Peter already built a few examples of templates containing data that the team can model to share as a prototype with Monica. "Monica has defined a base process unit template to get the team's feet wet," Bill said. "She created the templates from the data models used in the team's plant LP applications." Peter presented a series of illustrative examples to demonstrate how a new digital data infrastructure system would benefit ProcIndustries refinery operations.

Peter and Monica showed that they could configure a data object model to generate results for the entire refinery (Figure 2.4). They were able to present the object model that takes inventories, quality data, process data, and equipment data, and organize the data into a model they can manage to view operational metrics. The data associated with these variables are configured in the unit template in a standard way. The unit production targets are read by the system into a production table. The production table generates the time-interval events when the units are running OK, in trouble (such as below the production target), when the units are idle (no production but the equipment is available), when the unit is down (the equipment failed), or when the unit equipment is offline for scheduled maintenance.

Figure 2.5 illustrates how the digital data infrastructure system captures the refinery digital transformation team's knowledge about operations. In this example, the model gives Peter and his colleagues an illustration for everyone to reference—including executives at ProcIndustries who must buy into the digital data infrastructure project.

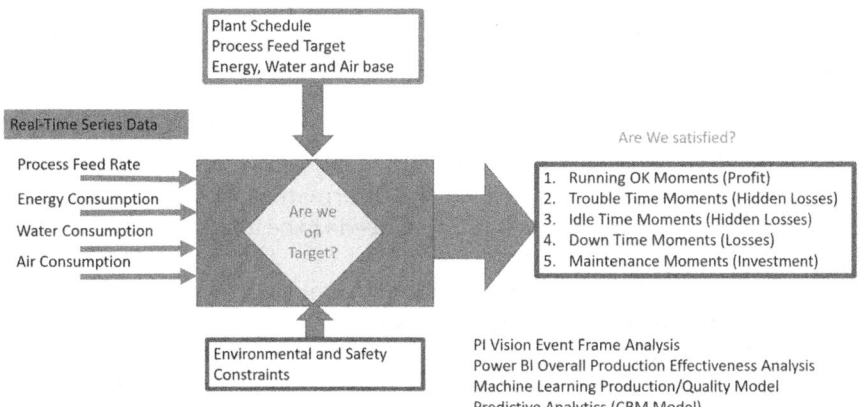

FIGURE 2.4
Smart process unit template schematic—condition-based maintenance (CBM).

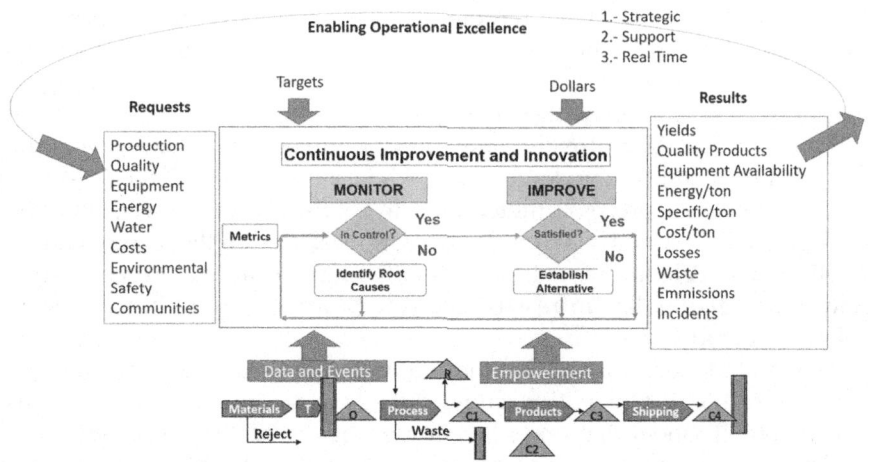

FIGURE 2.5
Generalized continuous improvement and innovation for an enterprise.

Operations Standardization for an Enterprise Competence Center

Tom, the refinery plant manager, said the presentation would give the digital transformation team an opportunity to gauge the reaction of his counterparts in other ProcIndustries refineries around the country. Most of the refineries have evolved over the years using the same, or similar, technologies as the refinery in South Texas. Each one has its own individually operated legacy system with accumulated local, siloed business practices.

Tom indicated that the current South Texas refinery ERP system has extensive transaction data including human resources, legal systems, financial systems, and materials and production planning, but the ERP system is not designed for supporting immediate operational decisions. For example, the ERP system contains total electricity data used for the month. However, that does not provide operations with any insight into the current daily electricity consumption for the refinery. It only contains the monthly amount to be paid.

Each plant is operated individually. To create a cohesive standard digital data infrastructure for collaboration, the information silos are the key problems to resolve. Eventually, each refinery needs to be integrated with the new operational excellence program.

Bill asked Tom to explain how a standard digital data infrastructure environment would enable the implementation of a continuous-improvement program for the entire company. Tom said Peter and the digital transformation team would explain at the company-wide meeting that they are working on a common naming convention for refineries, process unit areas, materials, and equipment so everyone across the company uses the same terms and sees a common usage of data when reporting and analyzing operations.

Building the Foundation

A common language will help integrate the data silos at each location. When installed, the real-time EIDI will connect all relevant systems and provide access to data from any asset.

Peter said it would be important to compare the company's current system to the new digital data infrastructure system ProcIndustries is implementing (Figure 2.6). Peter pulled up a diagram showing a typical architecture using an enterprise-wide digital data infrastructure rollout to an entire corporation, including multiple sites. Such a rollout is vital to maintain one version of the system, which supports data for many functions. It is a prerequisite to implementing an operational excellence program, which requires a standardized approach.

There are many benefits to standardizing a common real-time digital data infrastructure across multiple refineries. "Besides reducing implementation costs and standardizing system training, a common digital data infrastructure allows experienced engineers and managers to share their expertise and insights with colleagues in other refineries," Bill said. In addition, remote engineers and managers will easily be able to monitor overall fleet performance, while comparing metrics of similar units among several refineries.

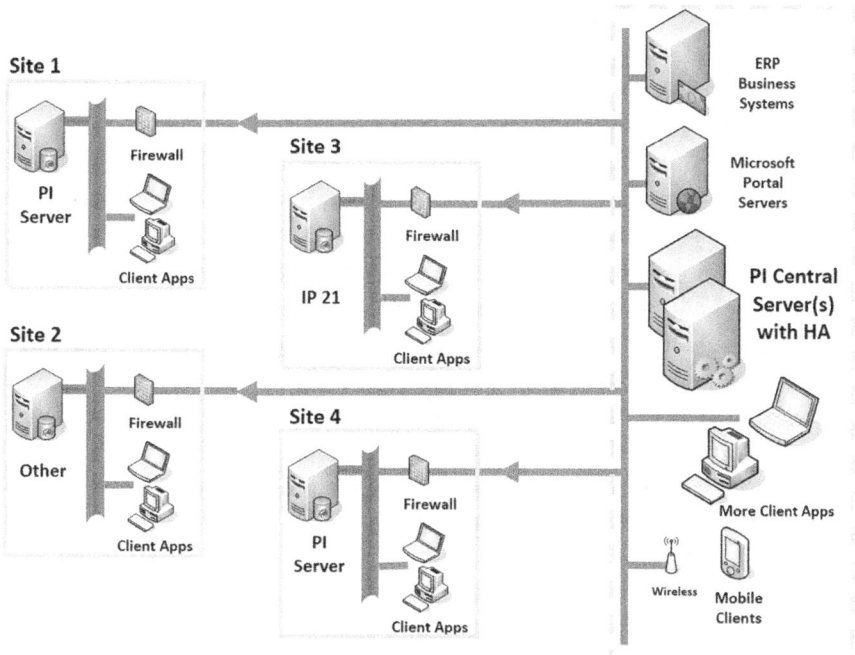

FIGURE 2.6
Enterprise data architecture across multiple plants.

Bill said the refinery sites can be located anywhere in the world. Figure 2.6 shows a schematic of the integration of many sites using a local digital data infrastructure connected to all areas at all of the company's refineries. *The system enables ProcIndustries to establish competence centers that provide assistance to remote refineries, which do not have enough staff members to perform real-time process analysis.* By working with in-house experts at a competence center, workers at a remote refinery can still participate in the continuous-improvement program.

Peter said that with advances in the industrial Internet of things (IIoT), cloud computing, mobility, and big data, the availability of sensor data from all types of assets will increase over time, ultimately expanding the digital data infrastructure system.

"The cloud provides a secure way to connect to distributed sites. Mobile computing devices, such as tablets, can connect people in real time. In addition, the digital data infrastructure can exploit big data and predictive analytics to improve the quality of the insights that our data analysis tools provide," said Peter.

As the meeting concluded, Peter summarized the benefits of the EIDI for different levels of the company:

Enterprise Level

- Enterprise functional and plant-level information is consistently based on the same version of data.
- Enterprise-wide real-time data access better leverages domain specialists for process support.
- Yield accounting is based on validated data, traceable to the source.
- Planning is based on real-time access to validated plant floor data on inventory and availability.
- Proactive condition-based maintenance management will be implemented and integrated with production planning and execution.
- Environmental reporting will integrate process operating data and environmental measurement systems.
- Safety-related data is integrated with process data and equipment data for risk analysis and assessment.
- The EIDI information can be extracted and integrated into corporate business systems and advanced analytics.

Regional Level

- Data sharing is enabled with higher level enterprise systems, which allows integration with content management, to support plant engineers with access to approved procedures, plant documentation, and current drawings.

Building the Foundation

- EIDI allows for standardization of process units for enterprise benchmarking and fleet-type enterprise evaluations for improvements.
- Real-time plant-level data collection is done with full auditing capability, with a trail to the source of the data.
- Material tracking and product genealogy functions will obtain EIDI data.
- This enterprise model enables real-time operational reporting and analysis on plant and process performance to all levels of management.
- Production volumes are based on real-time material balancing.
- Process yields are based on material balancing.
- Energy and water consumption are based in real-time balancing.

Individual Refineries

- The infrastructure architecture leverages existing data from distributed control systems (DCSs), programmable logic controllers (PLCs), and advanced process control models.
- It allows independent upgrades to plant-level automation systems (DCS, PLCs, etc.), as the EIDI is control-system agnostic.

"Competence centers are an important point to emphasize," said Peter. ProcIndustries plans to have competence centers to provide support and coaching for the whole enterprise. The competence centers will assist staff at the company's refineries to understand how to work with the digital data infrastructure and implement process improvements. For example, Peter said, new environmental guidelines require clear performance indicators for monitoring energy and water usage, as well as compliance with environmental and safety regulations at all sites.

To summarize, Bill shared the following illustration (Figure 2.7). With an employee at the center of the action, the digital data infrastructure provides information on key metrics so that an employee can perform his or her job, focus on ways to improve, ways to serve customers, and by doing so, improve the company's profitability. Additionally, it is well-known that when companies focus on the safety aspects, their profitability increases (Intech 2018).

Bill explained that it is ProcIndustries' vision to work on the six process improvement elements in tandem and be able to drill down to the desired level of detail. The system should provide easy access to operating procedures, safety and environmental regulations, asset mechanical information, quality guidelines, and chemical engineering process simulation.

"For example, during implementation of the U.S. Occupational Safety and Health Administration regulations for the oil industry, it became evident

FIGURE 2.7
The six elements to improve refinery operations.

that not everyone at ProcIndustries is familiar with the latest process safety-management information. I believe the digital data infrastructure will help us all get up to speed," he said (Bascur et al. 1992).

According to a survey by LNS Research, organizations using safety culture, procedures, and technologies avoid safety incidents and improve business performance. Organizations in which environmental health and safety (EHS), operations, and engineering collaborate to improve all aspects of safety report a median incident rate 15% lower than those without this tight collaboration. "Using this knowledge, our digital transformation team includes collaboration to increase safety as a viable part of their plan," noted Bill.

What You Should Take Away

After a significant incident, such as the refinery transformer explosion, it is best to start looking at current workflows and how data is used and shared by various teams, rather than simply rushing to implement new technology. The ProcIndustries digital transformation team analyzed all facets of refinery operations, including production, maintenance, safety, environmental, and utilities to develop a plan where the proposed EIDI information can be shared and effectively utilized by all workers. A key to success lies in management's ability to motivate people to buy in and use it daily to solve problems and improve the business. When embraced by people at all levels and implemented successfully, continuous improvement and innovation become the DNA of your organization.

Standardizing and building reusable process templates (objects) becomes the strategic new business model. This standardization enables an enterprise digital data infrastructure for business intelligence (BI), and facilitates predictive and proactive decision-making using real-time data variances (planned vs. actual results). The business operational excellence loop provides a mechanism for asking the right questions: Are we on target? Are we satisfied? If ProcIndustries succeeds at process and quality improvements, there will be opportunities for enterprise-wide innovation.

To accelerate the operational excellence program, a company benefits from appointing a coaching and support group. This special team maintains the business knowledge of the organization and offers the training required for the organization to adopt new best practices. Establishing domain-specific competence centers gives the company the means to provide coaching and knowledge that can be integrated with the whole business.

The ProcIndustries digital transformation team knows they need to change the organization's culture. The digital data infrastructure and applications to analyze data creates the opportunity to develop many small projects aimed at improving operations. These improvements can add up to reduce inefficiencies and operating costs. Nevertheless, the culture needs to be in place so that the people doing the work seize those opportunities.

The exchange of operational information, improving the collaboration among functions, will shrink the "white spaces" between them. Moreover, tracking the quality of these interactions requires operational data.

At ProcIndustries, a team of relative newcomers—Peter and Monica chief among them—was able to configure a model to analyze operations at their refinery, look at production and energy and water consumption of each of the refinery's process units, and suggest efforts that will yield opportunities.

References

Bascur, O.A., and Kennedy, J.P. 1995. "Measuring, managing and maximizing performance in petroleum refineries." *Journal of the Japanese Petroleum Institute* 102.

Bascur, O.A., Vogus, C.B., and Bosler, W.H. 1992. "Long-term knowledge integration with OSHA PSM." In *NPRA Computer Conference Proceedings,* November 16–18. Washington, DC: National Petroleum Refiners Association.

Duggan, K.J. 2012. *Design for Operational Excellence: A Breakthrough Strategy for Business Growth.* New York: McGraw-Hill.

Gonzalez, F. 2014. *Reinventing the Enterprise in the Digital Era.* Madrid, Spain: BBVA.

Harrington, H.J. 2012. *Streamlined Process Improvement.* New York: McGraw-Hill.

Intech. 2018. "Industrial companies using safety to increase profitability." *Intech Magazine* (November/December): 8.

Kennedy, J.P., and Bascur, O.A. 2002. "Influence of computer and information technology on process operations and business processes-a case study." In *Chemical*

Process Control VI: Assessment and New Directions for Research: Proceedings of the Sixth International Conference on Chemical Process Control, eds. J.B. Rawlings, B.A. Ogunnaike, and J.W. Eaton. CACHE Series, Vol. 98. New York: American Institute of Chemical Engineers (AICHE). pp. 7–11.

Miller, A. 2014. *Redefining Operational Excellence.* New York: American Management Association.

NIST (National Institute of Standards and Technology). 2017. "Malcolm Baldridge National Quality Award 1988 Recipient: Motorola Inc." www.nist.gov/sites/default/files/documents/2017/10/11/1988_Motorola_Inc.pdf. Accessed February 18, 2016. [July 20, 2019].

Rogers, D.L. 2016. *The Digital Playbook.* New York: Columbia University Press.

Rummler, G.A., and Brache, A.P. 2013. *Improving Performance: How to Manage the White Space on the Organization Chart,* 3rd ed. San Francisco, CA: Jossey-Bass.

Additional Reading

Arguelles, I. 2014. *New Work Places: Collaborative Work Impulse.* Spain: BBVA. pp. 295–334.

Bascur, O.A. 2019. "Process control and operational intelligence in mineral and metallurgical processing." In *SME Mineral Processing & Extractive Metallurgy Handbook,* vol. 1, ed. R.C. Dunne, S.K. Kawatra, and C.A. Young. Englewood, CO: Society for Mining, Metallurgy & Exploration.

Bascur, O.A., and Kennedy, J.P. 1996. "Measuring, managing, maximizing refinery performance." *Hydrocarbon Processing* 75(1):111–116.

Bascur, O.A., and O'Rourke, J. 2019. "Measuring, managing and transforming data for operational insights." In *Smart Manufacturing: Concepts and Methods,* ed. M. Soroush, M. Baldea, and T. Edgar. New York: Elsevier, forthcoming.

Bascur, O.A., and Soudek, A. 2019. "Grinding and flotation optimization using operational intelligence." *Mining, Metallurgy & Exploration* 36(1):139–149.

Brache, A.P., and Rummler, G.A. 1987. "Managing the white space on the organizational chart." Internal document, GTE Corporation.

Burke, T. 2018. "OPC Interoperability standard for industrial automation." *Intech Magazine* (November/December): 28–31.

Goldratt, E.M. 2014. *The Goal: A Process of Ongoing Improvement,* 4th rev. ed. Great Barrington, MA: North River Press.

Hales, K., and Lavery, M. 1991. *Workflow Management Software: The Business Opportunity.* London, UK: Ovum.

Harrington, H.J., Hoffherr, G.D., and Reid, R.P. 1998. *Statistical Analysis Simplified: An Easy-to-Understand Guide to SPC and Data Analysis.* New York: McGraw-Hill.

Hawkins, J. 2004. *On Intelligence.* New York: Henry Holt and Company.

Kotter, J.P. 2014. *XLR8 [Accelerate]: Building Strategic Agility for a Faster-Moving World.* Boston, MA: Harvard Business Review Press.

Maitland, A., and Thomson, P. 2014. *Future Work: Changing Organizational Culture for a New World of Work.* Basingstoke, UK: Palgrave Macmillan.

Markman, A. 2012. *Smart Thinking: Three Essential Keys to Solve Problems, Innovate and Get Things Done.* New York: Penguin.

Nicola, J., Mayfield, M., and Abney, M. 2002. *Streamlines Object Modeling.* Upper Saddle River, NJ: Prentice Hall.

Oakland, J.S. 1986. *Statistical Process Control.* New York: Wiley.

Pande, P.S., Neuman, R.P., and Cavanagh, R.R. 2000. *The Six Sigma Way.* New York: McGraw-Hill.

Plourde, M. 2016. "Heavy haul equipment logistic strategies in iron ore mining." Presented at the 2016 Users Conference, San Francisco, Process Industries, Transportation, and Supply Chain. www.osisoft.com/Presentations/Heavy-Haul-Equipment-Logistic-Strategies-in-Iron-Ore-Mining.

Plourde, M., Bascur, O.A., Paquet, S., and Gervais, D. 2017. "Digital innovation in modern engineering and operational excellence." *Presented at the 2017 SME Annual Conference and Expo,* Denver, February 19–22.

Plourde, M. 2019. "PI System at the heart of a mining integrated operations center (ArcelorMittal)." Presented at the 2019 PI World, San Francisco, Mining, Materials, and Supply Chain. www.osisoft.com/Presentations/The-PI-system-at-the-heart-of-a-Mining-Integrated-Operations-Center--ArcelorMittalx/.

Scholtes, P.R. 1988. *The Team Handbook: How to Use Teams to Improve Quality.* Madison, WI: Joiner Associates.

Steyn, J., Bascur, O.A., and Gorain, B. 2018. "Metallurgy analytics: Transforming plant data into actionable insights." *Mining Engineering* 70(9):18–29.

3

Using EIDI Data as a Strategic Asset

A small bunch of people who know what they are doing can accomplish more than a big group of people who don't know what they are doing.

Robert Noyce

Chapter Overview

In this chapter, the digital transformation team prioritizes refinery improvement needs and makes its first important decision, the selection of enterprise industrial data infrastructure (EIDI) software. We will describe what they hope to accomplish using the new EIDI and how they deploy it with a top-down approach. We will also outline the key EIDI capabilities and their importance to the refinery's business goals. Finally, we will describe how existing plant process diagrams assist in configuring the EIDI database.

The Need for an Enterprise Industrial Data Infrastructure

Peter Argus, continuous-improvement manager and leader of the digital transformation team, reflected on the team goal of digitizing the processes at the ProcIndustries' South Texas refinery and hopefully, throughout the enterprise.

At the latest transformation team meeting, the team recounted what transpired during the stormy afternoon when the electrical transformer blew out. Peter told the group, "No one expected the sudden explosion and fire. After the fact, we knew the refinery's staff was pushing the refinery and the equipment to their limits to meet production goals. This put too much burden on the transformer. At the time, it was not possible to see it—even though our refinery production manager, Tim Olsen, and others at the scene were working hard to gather information and make decisions."

"I thought a lot about what Tim had done in preparation for his decision-making," Peter said. Tim was collecting raw data from a variety of sources. He was organizing the data by location from the damaged and unharmed areas of the plant. He was performing his own analysis, mashing up the disparate data in his head to determine the major losses by area. Based on experience, he was creating a mental picture of exactly what was going on during the outage. Tim was able to communicate to the operators in the control room what they were seeing, so they, in turn, could arrange any necessary communications with the maintenance crews to assess the situation.

"Tim was doing everything he could," Peter said. However, without a comprehensive real-time data infrastructure, it took a great deal of manual effort to accomplish a fraction of what could have been done if they had their data infrastructure implemented. "Then, we'll be able to see unusual events like a power surge putting a lot of stress on the electrical system."

Several weeks after the incident, Peter brought together some of the transformation team members to meet with department team leaders at the refinery, specifically operations, maintenance, planning and economics, safety, and engineering. As he had hoped, the refinery team leaders shared their frustrations of not having enough key information readily available during an exceptional event.

As Peter later summarized the team leaders' experiences, he surmised the first and most important step was to combine and store as much refinery data as possible into a single location for immediate access when needed.

This requires several things, the first of which is to select and deploy a robust software system that reliably ingests and consolidates all refinery time-series data, establishing a single "system of record" for the refinery. This system of record creates a permanent data archive where all time-series data resides, including process data, equipment performance data, laboratory quality data, production data, utility data, energy and water consumption data, and so on. It is important for the data to be all in one place for real-time visibility, historical analysis, internal and external reporting, asset condition monitoring, and for ad hoc analysis, root-cause analysis, or whatever is needed to make refinery operations more efficient.

The team coined the term "enterprise industrial data infrastructure" or EIDI, as we refer to it throughout this book. The term *infrastructure* implies something that provides the capability for people to accomplish specific objectives over time, such as a company's information technology (IT) network or a country's system of highways or mass transit (Figure 3.1). An infrastructure is not something deployed for one specific purpose, or for a finite period; rather, it enables numerous things that people will employ over time. An infrastructure also facilitates unanticipated use cases to be implemented in the future.

Peter and the team would need to convince all refinery technical personnel to get their data from this single system of record, so that the data they used for their analysis would be accurate and easy to access. He knew that engineers had a difficult time getting data for their reports and analyses.

Using EIDI Data as a Strategic Asset

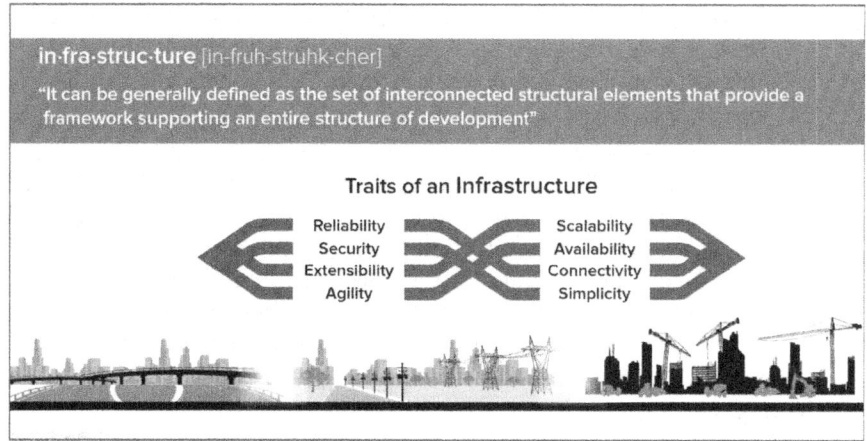

FIGURE 3.1
Definition and attributes of a real-time infrastructure.

They frequently had to import data from other people's spreadsheets and from databases hidden away or not easily accessible. Peter hoped refinery employees would be more highly motivated by using their time more effectively, by reducing wasted time searching for data and using that time to solve problems and improve refinery operations. He hoped that there would no longer be different versions of the same data, just one version: the truth.

Determining What Is Important

Peter and the team knew from previous experience that identifying needs and selecting the proper software were key activities they needed to execute correctly. The team worried a bit about not knowing enough and making a wrong decision in this regard, but they visited with several industry leaders in the oil and gas and power-generation industries. Upon completion of their diligence meetings, they decided to prioritize what they needed into three distinct categories: company (vendor) selection, current needs, and possible future needs. Furthermore, in each category, they listed the most important requirements:

1. **Infrastructure software supplier**
    - Deploy the best possible software system, the best in its class, which has been used by other leading energy and manufacturing companies. The software should have a distinguished history of use by leading energy producers and production companies.

- Have reliable 24/7/365 support for the software. The proposed EIDI is to be mission critical for ProcIndustries operations, as the refinery continuously makes product.
- Restrict the list of suppliers to firms that are independent, not an extension of a hardware or control system company, as the refinery needs to collect and consolidate data from various control system manufacturers. The EIDI will ingest data from other diverse systems such as laboratory systems, a utilities SCADA (supervisory control and data acquisition) system, the ProcIndustries production scheduling system, and planned Internet of things (IoT) sensors, needed to augment instrumentation. These IoT sensors could potentially be supplied by anyone.

2. **Current needs**
   - Integrate all available digital real-time data from disparate systems into a common system of record. This should include existing legacy systems. Data should be time-stamped and synchronized with current process conditions, unless there are specific needs to do otherwise. Data losses from lost connections should be less than 1%.
   - Ensure all data is permanently stored in a reliable archive. Data stored into the archive must not be lost and must be quickly and easily retrievable for trending, reports (typically via Microsoft Excel spreadsheets and web-based team sites), displays, historical analysis, and when performing ad hoc queries.
   - Employees should be able to quickly and easily visualize EIDI information via role-based dashboards, displays, and reports. End users need to be able to create and modify these displays and reports, as necessary for their job requirements.
   - The EIDI should be able to calculate derived information, so that calculation results are available for everyone to access. These calculations should not be done in Excel spreadsheets.
   - Have the ability to extract historically archived information for use with other software systems (e.g., production scheduling, process models, corporate web portals, analytic tools) without undue effort.
   - Be able to assign physical context to data that is collected. With the cross section of teams needing this information, it must be intuitive to understand what the data represents. For example, distillation column tray temperature no. 3 is more useful than having to remember data variable name "TI403." A hierarchical data model that translates data to the physical representation of the plant is essential.

Using EIDI Data as a Strategic Asset 65

- The EIDI system should make it easy for personnel to access information from specific production events, such as a specific grade of crude or a specific product run. The data would automatically be framed and time-sliced for that event.
- The EIDI should be able to be configured within a reasonable period of time. Initially, ProcIndustries would need deployment assistance from the software provider, or skilled systems integrator with EIDI experience, or internal IT resources. However, after the initial deployment phase, the EIDI should not require continued use of external resources for normal EIDI maintenance. The software supplier should provide reusable, template-based configuration tools for easier, quicker system deployment.

3. **Future needs**
    - The EIDI should be sufficiently scalable to perform the preceding activities and to be capable of storing and managing additional data required for future use cases, as well as be able to scale to accommodate an increase in the number of users accessing the system.
    - The EIDI must be capable of providing data to ProcIndustries business and technical applications, either within the company's IT network or in a cloud environment.
    - There should be standard mechanisms so that employees in non-refinery areas (e.g., business analysts and data scientists) can obtain EIDI-stored information for input to business systems, corporate reports, engineering/process models, and cloud-based analytics.
    - In specific cases, ProcIndustries may need to provide EIDI data to external entities. There should be a method to accomplish that.

What Is an Enterprise Industrial Data Infrastructure?

Peter noted there are three basic EIDI capabilities that will be critical to the South Texas refinery: *data management, analysis,* and *data visibility*. He proceeded to summarize each category:

1. *Data management* involves the ingress, management, and storage of data associated with refinery assets so that the correct information is available for everyone in the context they need. Effective

data management requires sophisticated interfacing to all real-time data systems in the refinery. These subsystems include sensors, equipment measurements, distributed control systems, utility SCADA systems (for energy, water, and steam), equipment performance, product quality results from laboratory information systems, inventories, environmental measurements, and weather-related data. The data is stored in perpetuity so that everyone in the company can access the data they need at any time going forward.

The EIDI contains an object-oriented data model that represents refinery assets in hierarchal form and drills down to production lines, process units, and even individual equipment. This asset data model, commonly known as an *asset framework*, makes it easier for system users to understand what data they are seeing, where it is, and how it relates to the refinery process. Peter said that by organizing information by process unit, it will make it easier for Tim, the production manager, to make quicker business decisions, for example, about to where to send maintenance crews to repair damaged equipment during storms and electrical outages.

In addition, the EIDI can aggregate data into information using framed time intervals that represent specific production events. This will help compare similar plant events, such as unit start-ups, running a specific grade of fuel, or making a specific product. Assigning physical and event-related context makes it much easier for producing reports and displays, as well as facilitating business and analytic tools, such as machine learning (ML).

2. *Analysis* represents the ability of the EIDI to determine conditions, excursions, and results much faster than human capability. Specifically, the EIDI assesses what is going on right now and which area of the refinery or process unit needs attention. The EIDI also highlights where there are opportunities for improvement, providing the supporting data at the desired detail needed for analysis and decision-making.

The EIDI accomplishes much of this through real-time calculations, with calculation results also known as *derived* or *inferred variables*. Derived variables can be common statistical measurements or custom calculations defined by the user. These results are permanently stored into the EIDI archive, as are other measured data coming from sensors via plant control systems. If the EIDI detects that a sensor measurement or a calculated result is out of acceptable range (as configured by the user), the EIDI has the capability to transmit real-time notification alerts, indicating a condition that needs attention. Peter remarked to the group: "If the refinery had this capability, we may have averted the transformer catastrophe."

By effectively managing the data, placing it into proper context, using event frames, and executing real-time calculations, the EIDI facilitates basic root-cause trending analysis up to more sophisticated user queries such as the following:
- What is the feed rate for Unit XYZ in Area 1?
- What is the total amount of time that Unit XYZ has been in shutdown mode, in idle, on target, or in a trouble mode?
- What was the total amount of electricity consumed while Unit XYZ was idle?
- What have the cycle times been for producing Product A?
- What was the maximum flow rate achieved during the last work shift for Unit XYZ in Area 2?
- How many times has Unit XYZ been in an idle state during the last month?

In addition to the EIDI's online analysis capabilities, the EIDI can export large quantities of high-fidelity contextualized data and publish it as a data set for use by non-EIDI analytic systems, such as advanced visualization systems and predictive modeling tools. We explore this concept in Chapter 10.

3. *Data visibility* is essential to have a window into the refining process for real-time situational awareness and for historical analysis and reporting. Role-based dashboards and displays are key so that people with specific roles get the information they need in context of their assigned duties. For example, a maintenance engineer might have a list of the worst performing assets in a particular unit or the equipment uptime of that engineer's area over a given time period. Alternatively, a process engineer is less equipment focused, but is concerned about how effective and efficient the process is for a particular area or unit in the refinery. A production manager is interested in actual production rates versus targets, how costs compare to refinery standards, and how much energy is consumed in a certain area. To summarize, it is key to provide information that is relevant to knowledge workers. If needed, they can drill down or query other areas.
    - Role-based key performance indicator (KPI) dashboards, displays, and reports must be easily consumed and configurable by end users. This becomes part of their job. They are able to make better decisions.
    - The system must offer mobility capability for authorized knowledge workers to access data anywhere via their mobile phones, tablets, or laptops.

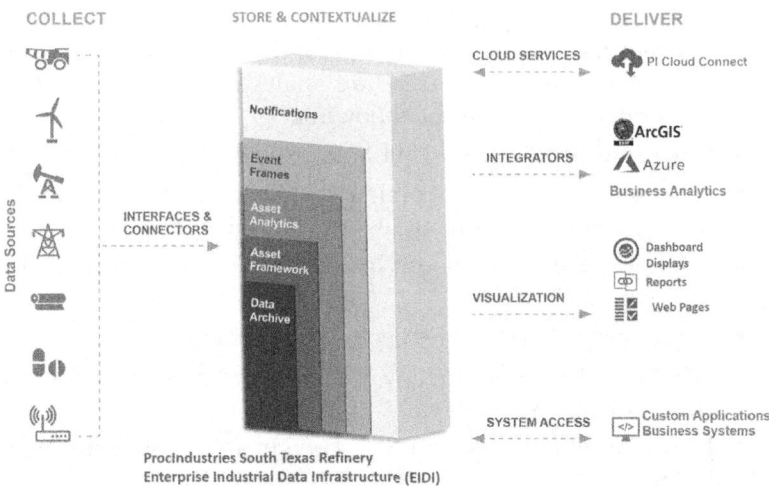

FIGURE 3.2
EIDI functionality, data ingress, and egress.

Figure 3.2 visually represents the preceding summary of EIDI functionality and how data is collected from sensors and plant control systems. It also highlights the various ways people and software access the EIDI information.

A Decision Is Made

The team had to make a decision about which software system to purchase, as management pushed them to start rectifying the situation so that a similar problem did not reoccur. Peter had experience with real-time infrastructure software in the past, but decided to seek out people in refining, in petrochemical plants, and in power-generation companies who had gone through a similar initiative and made a similar decision. The group scheduled visits with these companies.

The team found those sessions extremely constructive in that these companies confirmed that Peter and the team had defined the right selection criteria and had more or less prioritized them as these other companies had done during their selection process. They took away some things they had not anticipated, as these companies were honest about what they had done correctly and incorrectly during the deployment process. The first thing they usually mentioned was that the EIDI should be the official system of record and the refinery should not duplicate data in other places unless absolutely necessary because of compliance mandates, as in collecting emissions data, or because of some ProcIndustries' specific requirement.

They gave advice on items such as making sure end users have a voice and buy-in on how they wish to view and utilize the data. They also provided some guidance on how to train their EIDI users effectively on the new system and some guidance on how to communicate progress with upper management and temper their expectations.

One of the biggest takeaways was that these advising companies counseled ProcIndustries not to initially shoot for the moon with complex applications that would delay the refinery using the EIDI. Rather, they should look to find quick wins from deploying the EIDI using its standard functionality, not waiting to deploy additional software applications during the initial phase.

By adding some simple performance metric calculations and generating real-time alerts to notify refinery personnel when those results deviated from acceptable ranges, they could quickly catch things such as the following:

- They could avoid an unscheduled outage for a pump, motor, condenser, heat exchanger, or other critical piece of equipment that would risk a unit shutdown.
- They could make quicker discoveries when performing root-cause analysis through use of the EIDI's historical trend analysis capabilities and annotating in the historical data what caused the problem. This was very useful when trying to analyze why yield, throughput, or product quality was inconsistent.
- They could measure energy usage and confirm which assets were causing excess consumption.

The other takeaway was that these companies insisted that they quantify these quick wins for management and others to see. For example, if a pump shutdown was averted, the savings should be documented as the cost of a pump repair or replacement plus labor costs to repair/replace the pump, in addition to the lost revenue from not making product during the pump shutdown.

After meeting with the software vendors, viewing demonstrations, sifting through pricing proposals, and reviewing the decision with business and IT managers, the decision was made to use the OSIsoft PI System as ProcIndustries' new EIDI.

A Modern EIDI Implementation Approach

Now that the team was ready to go, the first step in the deployment process was to meet with the vendor and the ProcIndustries IT team to determine an appropriate architecture strategy that supported usage of the EIDI and its permanent data archive. The architecture also needed to accommodate the many knowledge workers who would consistently access data once they started using the EIDI in their daily routine. Fortunately, the EIDI had a flexible architecture that allowed server load balancing and provided a hardened method of deploying patches and updates, without unduly impacting EIDI server uptime.

IT also needed to anticipate that many more ProcIndustries people from various departments would begin querying the data once they learned of its capabilities. They decided to revisit the proposed architecture plan in six months to accommodate future growth. This growth would be likely because a significant number of companies now use more advanced modeling and predictive data analysis technologies, such as artificial intelligence (AI), ML, big data analysis, and business intelligence (BI) tools. All of these new technologies have one common denominator: They require vast amounts of accurate high-fidelity data, which the EIDI would supply.

As the suite of software and control systems used by an enterprise constantly expands, it is necessary to architect such systems in secure environments that take advantage of

- The latest and most efficient hardware offerings for processing, memory, and storage;
- Fast, reliable networks that are able to transmit large amounts of data needed by people or software systems to analyze information; and
- The latest and most effective cybersecurity practices and software tools to prevent unauthorized people or software from accessing EIDI information. Equally as important, no one except for authorized plant operations and IT personnel should have access to plant control networks. Typically, customers place an EIDI on the company business network, where large data queries do not compromise the execution of plant control and safety systems. It also keeps nonessential operating personnel outside of the plant or refinery control network.

Figure 3.3 presents a proven EIDI deployment for a large industrial company, integrating process control systems with an operational management infrastructure. It is composed of

- Specialized data connectors and interfaces to specialized control and equipment systems, weather systems, laboratory systems, text files, web interfaces, and geographical information systems;
- Real-time data contextualization, visualization, online analysis, and real-time alerting;
- Operational data modeling, BI, ML, and process advisory tools; and
- A demilitarized zone representing a cybersecurity strategy of separating business and control networks.

Using EIDI Data as a Strategic Asset

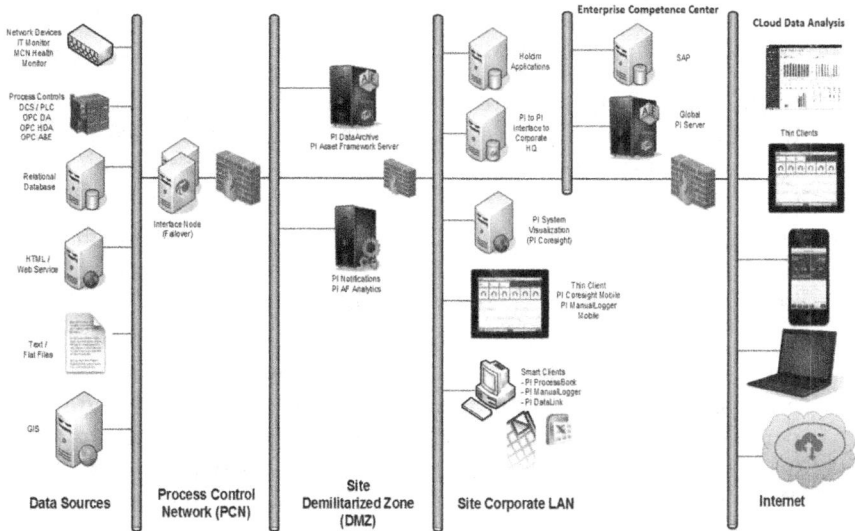

FIGURE 3.3
Proven EIDI architecture for large enterprises. (Courtesy of OSIsoft.)

The left side of the diagram shows the types of EIDI connectors and interfaces to ingest and store refinery data in control systems: sensors, laboratory systems, along with software or web-based information, and IoT edge devices.

Companies can leverage their existing investments in legacy data collection systems. They are able to integrate all silos of data without having to replace their existing legacy systems (Kennedy 2019). A robust EIDI should be hardware and control system vendor independent. This homogeneous data environment brings new opportunities, such as managing and analyzing operational data from a business perspective (Logue 2019). The existence of a real-time data infrastructure (EIDI) encourages curiosity and innovation, which continuously increases business value.

Decision-Making through Plant Data Hierarchy

Peter knew from his discussions with the advising companies that it was critical to begin EIDI deployment by mapping sensor data sent from the refinery's plant control systems (distributed control systems [DCSs], programmable logic controllers [PLCs], SCADA, and IoT devices) to a physical or process entity that everyone understood. The EIDI database needed to be represented as a hierarchical data model that represented the refinery, its units, and its equipment. There are three advantages to doing it this way:

1. Users know what a data stream represents, for example, crude unit feed-water temperature or pump number 2 intake pressure, as opposed to some instrument nomenclature that are only known by a few people.
2. Humans and software systems are able to consume information represented as an asset, comprised of numerous data points, versus having to search for the individual data points that comprise the asset.
3. These assets are defined in the EIDI through configurable reusable asset templates. This means that for the many heat exchangers, pumps, or distillation column tray temperatures, one single template per asset type is defined and easily repopulated. This eliminates the need to configure each asset anew and reduces deployment time. These templates consist of:
   - Representing the data points that comprise a particular asset;
   - Defining an inline calculation to provide a derived value for that asset (e.g., degrees Celsius to degrees Fahrenheit, or heat exchanger fouling value);
   - Triggering a real-time alert if a specific metric or value falls outside of acceptable operating ranges; and
   - Including other non-time-series-related data, such as the last time the asset was serviced or the maximum allowable rate at which the asset can be run.

Figure 3.4 displays the components of an asset data template.

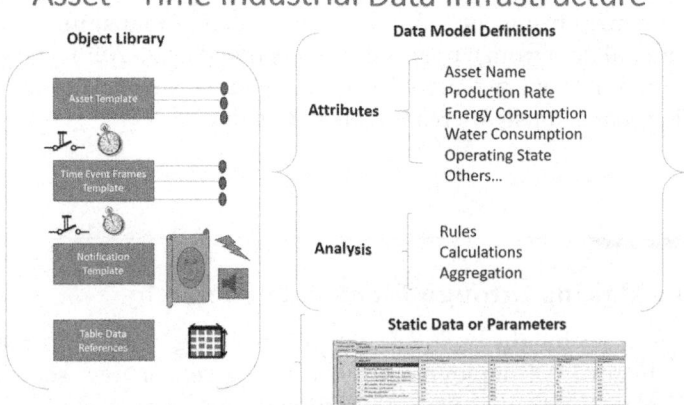

FIGURE 3.4
Asset framework object: template components.

EIDI Deployment and Configuring the EIDI Templates

The team began undertaking EIDI deployment. They successfully achieved the first milestone, which was to ingest and consolidate refinery data from the control systems into the EIDI, creating a duplicate of essential refinery information in a safe analysis environment. This information is now accessible by knowledge workers. Refinery personnel no longer had to search for data on the control system network or try to find it in some hidden silo that took valuable time to collect. The team had built several master versions of dashboards and displays, with the understanding that these standard displays would enable people to start analyzing data quickly. There were a few old-timers who initially resisted, but soon thereafter, every employee realized the refinery had to change, and most everyone got on board and used the EIDI.

Using Templates to Define Assets and Event Frames

The basic EIDI asset framework provides the capability to represent the refinery data in a hierarchical object-oriented schema. The ProcIndustries' team needs to define asset templates and reuse them for similar types of equipment to achieve the greatest impact of the real-time data infrastructure. They also need to use *event frames* and combine the two EIDI capabilities.

Once refinery personnel started adopting and using the EIDI data, the team turned their attention to designing the hierarchical data model and defining the most commonly used asset templates. Peter assembled the group to determine how best to tackle this important issue, which would have lasting consequences for the refinery.

The team would use another capability of these object asset templates; they would be used in the context of defined production or refinery events. For example, if there were similar events that needed to be analyzed comparatively, the EIDI could segment and time-slice the continuously collected data so that ProcIndustries personnel could compare and analyze similar events, such as producing gasoline using a specific type of crude oil or noting unit startups or shutdown events. Figure 3.5 shows an asset template with event frame context.

One of the most resourceful and productive methods of using these event frames is to dynamically create events, based on the process or operating status of the unit. This approach is used to create a "trouble" event when a unit is in an underperforming or a trouble operating mode.

This means that something caused the unit to run outside of acceptable conditions. Because the EIDI or operations personnel don't generally know when a unit enters a trouble mode, the EIDI can dynamically create such an event type during or after unit operation. When a metric or value deviates from normal conditions, the EIDI can trigger a start to the event. Consequently,

Asset Data and Time Context Framework

Process Unit Template

- Library
  - Asset Template
  - Time & EVENT
- Attributes Definitions
  - Production Rate
  - Energy Consumption
  - Water Consumption
- Analysis Type: Rules, Calculations
- Expression
- Event Frames Generation
- Rollup: $\int_{St}^{Et} Attribute(t)\, dt$

FIGURE 3.5
Asset template object components with the capability for event context.

if or when the unit returns to normal operation, the EIDI can trigger an end-of-event indicator. The EIDI even has the capability for people to mark these events after the production run has completed if it has been determined that there were problems. This last capability is powerful but should only be activated by people who are very familiar with the process and the equipment, until those people can determine how to automate the event triggers.

Figures 3.6 illustrates this dynamic time-slicing capability.

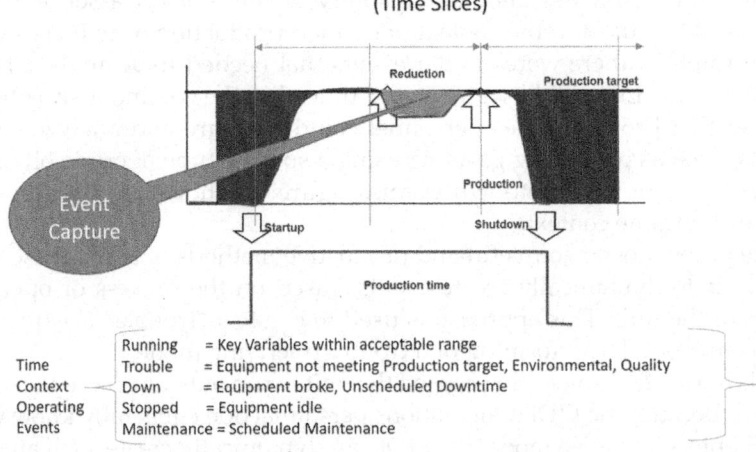

Process Unit Operating Events
(Time Slices)

Event Capture

Time Context — Operating Events

Running	= Key Variables within acceptable range
Trouble	= Equipment not meeting Production target, Environmental, Quality
Down	= Equipment broke, Unscheduled Downtime
Stopped	= Equipment idle
Maintenance	= Scheduled Maintenance

FIGURE 3.6
Defining operating modes for a process unit.

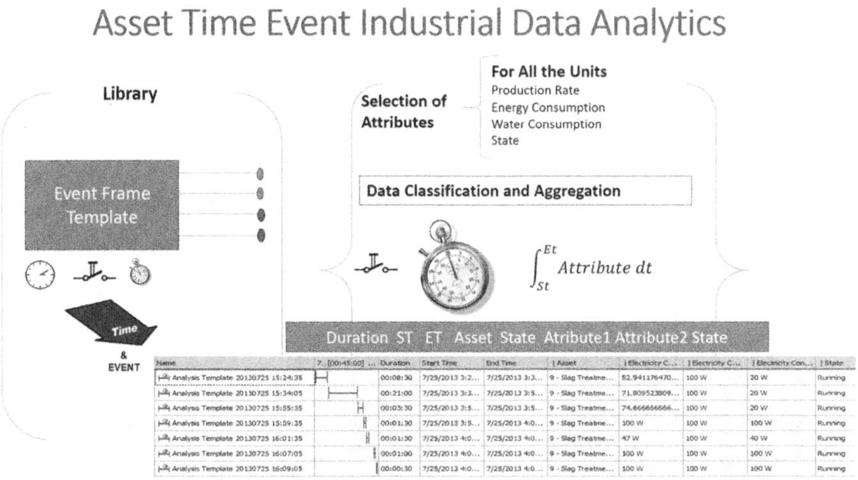

Using EIDI Data as a Strategic Asset 75

Asset Time Event Industrial Data Analytics

FIGURE 3.7
Event frame template configuration and the event-framed data results.

Peter discussed the implications of these capabilities: "The time-event context stores the start and end times to calculate an asset's time-derived variables over a defined period of time. This will be key for analysis and optimization. As such, we can store totals and averages for selected attributes, calculated at the defined operating start and end times. This transforms the real-time and statistical data into operational intelligent information." Figure 3.7 shows how the event frame capability creates the resulting event-framed real-time data table. This information is analyzed directly through EIDI visualization tools or it can be exported to external products, such as Microsoft Excel or Power BI.

Data Object Models

Peter displayed a simplified diagram that shows a typical operational data hierarchy for capturing, storing, and tailoring the data for analysis by people at many levels of an organization (Figure 3.8). The big difference from traditional data-gathering systems, which do not use time-series algorithms to manage data at the original resolution, or simple data historians, is the data hierarchy. The hierarchy arranges and stratifies the raw streaming (real-time) data into the desired level of detail, which is available on demand by other systems or users.

The data is reusable, starting from the plant process-control level through to the enterprise level. "Data classification, the process of categorizing data

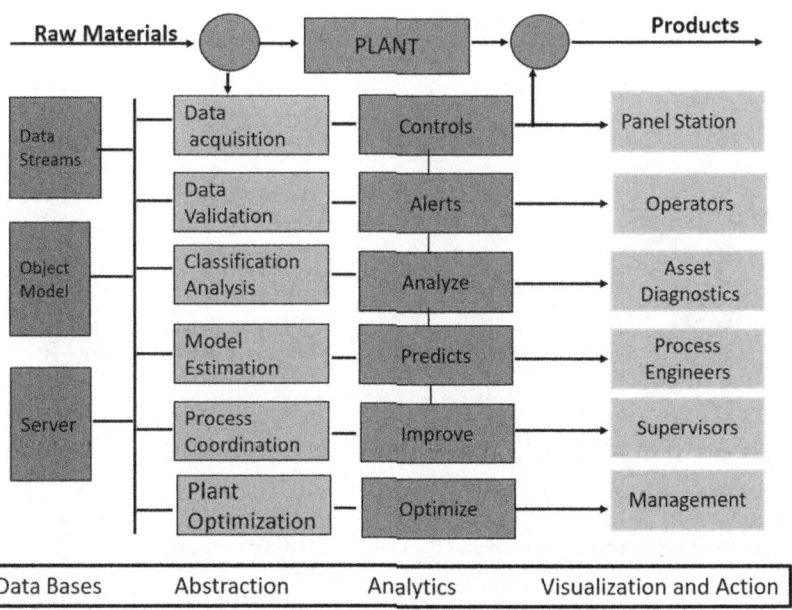

FIGURE 3.8
ProcIndustries' data transformation via layered analytics, shown by role.

to make it useful, is an important, but often ignored step in the analysis," Peter said. "The operational states of different assets are defined, so that we can analyze the data describing those states. Then, we can evaluate how the process units are performing. We can assess their operating costs from consumable variables such as energy, water, and so forth. Once the unit operational state is available, it becomes possible to perform quality and efficiency calculations to evaluate process performance, and ultimately, find ways to optimize the process" (Bascur 1988, 1999; Bascur and Halhead 2013).

Once the state of the process units is available for everyone to see, the next level of analysis is to evaluate whether refinery production is meeting company goals. "Then we can identify ways to optimize refinery production scheduling, match supply to customer demand, and increase overall plant capacity."

Peter explained that the object model database enables the EIDI to add context to the data based on assets and operating conditions of those assets, corresponding to the time intervals (or events) stored in the data archive. The time intervals facilitate data aggregation at the right time intervals and the right degree of detail. "It is the key to transforming the data into meaningful information!" Peter exclaimed.

With the EIDI, ProcIndustries can integrate the refinery data with other relational databases, shown as "Server" in the "Databases" column in the diagram. The server might have all the process diagrams and documentation, process safety information, product quality specifications, and production planning targets.

The "Abstraction" column is the data hierarchy column and shows how the EIDI transforms data at each level of decision-making. The "Analytics" column shows the process analytics or real-time calculations. The "Visualization and Action" column represents role-based data visualizations.

Data Acquisition, Validation, and Classification

"The data hierarchy shows how data is processed, transformed, and used at many levels of decision-making based on the context of time and other attributes associated with it," asserted Peter. "Once you have the right time context and operating conditions, you can apply a process model." Peter explained further, "You can calculate the efficiency of a process or predict a sensor reading."

The first step is to automatically collect data from all possible sources. The second step checks the connectivity of the subsystems and the validity of the data based on the basic limits of the sensors. The next step is *data classification*: Here an online analytics application checks the state of the process equipment based on basic performance rules. The state is stored as an event marker in time. This internal record of time is used to

- Aggregate all the data associated with a corresponding asset or physical entity;
- Establish a state that can be used to generate a notification to inform people about important updates; and
- Use the established state to aggregate more data.

Take, for example, a storage tank. A large tank will have a very long residence time depending on its size and the velocity of the pump associated with it. The pump's status is checked every one or two minutes. These status checks create a stream of data collected in the EIDI for analysis of what is happening and when with equipment and processes in the refinery.

"This step represents the 'Are we on target?' assessment we discussed earlier," acknowledged Peter. "Using this type of data classification enables us to sort the data so we can use it for additional value-added calculations. It augments the engineering knowledge available from the process engineers. For example, if it is within the reasonable operational limits, we can use the data in several ways: calculating performance, estimating the fouling factor of a heat exchanger, or evaluating the property of a stream using chemical engineering principles." This idea emanates from Bill Roberts' vision of automating to achieve continuous improvement and innovation, and notifying people immediately. The vision becomes a reality by automating the system monitoring "Are we on target?" loop and generating events and notifications (Bascur and Soudek 2019).

Block and Process Flow Diagrams

Block flow diagrams and process flow diagrams are familiar tools to engineers. They are important in digitizing industrial process plants. Process engineers learn about the block flow diagram early in their engineering education. The *block flow diagram* is a series of blocks connected with input and output flow streams (Turton et al. 1998, 2018; Woods 2005). The block flow diagram includes operating conditions (e.g., flow, temperature, and pressure) and other important information, such as conversion and recovery. It does not provide details regarding what is involved with the blocks, but concentrates on the main flow of the process.

Block Flow Diagrams

According to Turton et al. (2018), the block flow diagram can take one of two forms: a single process or a complete plant complex involving many different processes. The block flow diagram, as described in Table 3.1 and illustrated in Figure 3.9, represents a single process.

Figure 3.9 shows a simplified block flow diagram for the ProcIndustries South Texas oil refinery. Each block in this diagram represents a complete chemical process. If further detail is needed, a block diagram can be drawn for each of the chemical process blocks.

Peter explained, "The advantage of such a diagram is that we get a complete picture of what the South Texas refinery does. And we can see the major

TABLE 3.1

Block Flow Diagram Conventions and Format

Block Flow Diagram Construction
1. Operations are shown by blocks.
2. Major flow lines are shown with arrows giving the direction of the flow.
3. The flow goes from left to right whenever possible.
4. Light stream (gases) are shown toward the top with heavy stream (liquids and solids) toward the bottom.
5. Critical information unique to the process is supplied.
6. If lines cross, then the horizontal line is continuous, and the vertical line is broken.
7. A simplified material balance is provided.

Using EIDI Data as a Strategic Asset

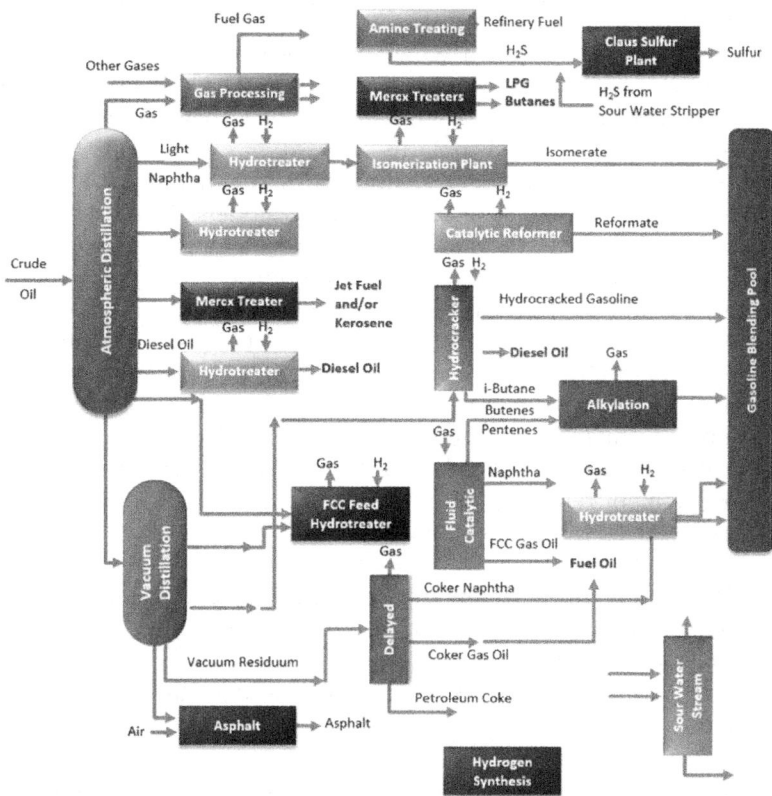

FIGURE 3.9
South Texas refinery process block diagram.

areas that need to be digitized, including all the important flows of material and information. The diagram includes associated sensors to incorporate into the EIDI context object model."

The block diagram also provides information on how all the different process units interact. This information is vital for analyzing interactions among the process units and to identify the response times to flow, temperature, and composition changes. To keep the diagram relatively uncluttered, only limited information is available about each process unit.

Process Flow Diagrams

A *process flow diagram* is used for a plant with many processes. These diagrams contain the basic plant topology and the connections to model the data into a production plant model. Peter instructed the team that they would be using process flow diagrams to build the context necessary for EIDI data.

"To gather the required information for each of the process units, we need to build a process flow diagram, which includes references to the sensor tags or identifiers from the control system," said Peter. The process flow diagram represents a quantum step up from the block flow diagram in terms of the amount of information that it contains.

Peter explained, "Process flow diagrams describe the physical plant, and we can use them to develop process simulation models." These diagrams are usually developed when the plant is constructed, and they are similar to the architectural drawings used when building a house.

For the South Texas refinery, the process flow diagram contains the bulk of the process and maintenance engineering data necessary for the design of a metallurgical or chemical process. In addition, process engineers use the process flow diagram to abstract events in the plant.

For plant information management, the objective is to capture the necessary information for performance monitoring of the process (production quality, equipment reliability, and environmental monitoring). The process information is used to predict quality variables or to define when the system should alert engineers to assets that fall below quality levels set in target plans. "While most process flow diagrams include similar information, the process flow diagrams from one company vary from those describing the same process at another company," offered Peter.

A typical commercial process flow diagram contains the following information:

- All the major pieces of equipment in the process are represented on the diagram along with a description of the equipment. Each piece of equipment is assigned a unique equipment number and a descriptive name.
- All process flow streams are shown and identified by a number. A description of the process conditions and chemical compositions of each stream are included. These data are displayed either directly on the process flow diagram or included in an accompanying flow summary table.
- All utility streams supplied to major processing equipment are shown. Each piece of equipment has its own description of the process variable that should be included. The diagram also should note the unique sensor tags associated with each piece of equipment. Doing this provides information to build the contextual asset object model and the associated physical description.
- Basic control loops are displayed, illustrating the control strategy used to operate the process unit during normal operations.

The same principles used to draw block flow or process flow diagrams can be used to develop powerful information linked to real-time and historical data. The elements making up the process flow diagram are created from templates whose additional properties and attributes can have data references to external databases. For example, the template attributes can store the equipment numbers and physical descriptions of equipment.

Peter announced that he wanted the digital transformation team to use process flow diagrams to start organizing the data generated in the process units (a process unit is composed of several pieces of equipment and components and includes details about individual devices or pieces of equipment).

Peter said that Monica Armstrong (planning and economics coordinator) could build on the foundation established by the process flow diagrams. She offered to provide a plant unit block flow diagram to organize the basic data necessary to integrate requirements for the plant schedule she prepares daily. This would add data definitions for products, maintenance, quality, and inventory. Combining these elements will make it possible to report the performance of all units based on a daily schedule.

Peter observed, "The process flow diagram provides all of the information we need to establish process-control protocols and prepare cost estimates to determine economic viability for the continuous process improvements. Using process flow diagrams with real-time data and events will help create processing plant optimization."

"Process flow diagrams are a key ingredient in digitizing the refinery," explained Peter. "In conjunction with the EIDI, the process flow diagram is used to proactively estimate equipment conditions and resources used in the refinery. It can help pinpoint issues, such as the efficiency of the pumps and compressors. It can be used for preparing estimates for inventories, and by using online process analytics, it can be used for predicting product quality."

Peter continued, "In addition, it is the basis for developing process simulators for mass and heat balances. It is also regularly used to diagnose operating problems that arise and to predict effects of process changes. Moreover, the process flow diagram is used to train operators and new engineers" (Bascur 2019; Steyn et al. 2018).

Organizing Operational Data

When working with an EIDI system, it is important to include materials and energy flows in the process plant as well as the data related to the fundamental process transformation that goes from raw materials to products. The EIDI handles the flow of information as the processes are transforming raw materials into semi-finished and final products for sale.

ProcIndustries uses a basic structure to standardize navigation and access to data in the refinery and will be using this structure throughout the enterprise as they expand use of the EIDI. Table 3.2 is a hierarchal model of ProcIndustries beginning with enterprise, then site, plant, plant area, unit, equipment, and ending with devices.

TABLE 3.2

ProcIndustries Physical Model

Physical Model	Abstraction	Typical Model
Enterprise	A collection of sites	A block flow diagram of a plant including inventories, receivables, shipping, enterprise resource management (ERP) software, key performance indicators (KPIs), and production rollup data.
Refinery (site)	A set of plants within a refinery, including inventories, receivables, and shipping.	A plant layout map, KPIs, total production, and yields.
Plant	A set of processing units within a plant area including inventories, receivables, and shipping.	Block flow diagram, KPIs, production, and performance information.
Plant area	A set of units	Process flow diagram, production, and performance information.
Unit	Unit with all sensors defined	Process flow diagram, production, performance, and quality.
Equipment	Equipment with all sensors included	Process flow diagram for equipment, real-time condition monitoring. Run-time performance calculations (run times, stop/start counts), manufacturing name plate, date of last/next planned service, work management (dollars spent, dollars planned to be spent).
Device	Detailed sensor locations	Piping and instrumentation diagram.

Process Flow Diagram Leads to Digitizing the Plant

"These diagrams and hierarchies serve as foundational information requirements to digitize the plant," Peter said. "We can model the data from the process diagram using a unit data model and configuring all of its attributes, analytics, and event generation at the object level." As such, the real-time unstructured data becomes information for people at all levels of the organization for real-time collaboration (see Chapter 5) (Steyn et al. 2018).

What You Should Take Away

The EIDI collects raw time-series data from many sensors, control systems, and software systems to integrate real-time data and relevant operational information into a permanent data archive. This data infrastructure becomes the operations system of record and is used by humans and other software systems for specific functions, such as process analysis, production reporting, and business system integration.

The ProcIndustries South Texas refinery digital transformation team prioritized what outcomes were important to getting the refinery to adopt best operating practices:

1. Aggregate all digitized real-time information to consolidate siloed data into a time-series data system of record.
2. Institute real-time situational awareness to identify and correct operational, safety, and maintenance problems before they become serious.
3. Continuously improve asset reliability to reduce unscheduled downtime.
4. Identify and correct bottlenecks in production and processes.
5. Manage energy consumption and water usage.
6. Create the ability to produce scheduled and ad hoc reports for internal consumption and for external compliance.
7. Generate real-time alerts when equipment or operating conditions deviate from accepted ranges.
8. Establish the ability to send data and interface with other ProcIndustries software systems, such as business systems, laboratory quality systems, and emissions monitoring systems.

To achieve this, they have identified key EIDI functions that enable them to reach these goals:

- The ability to collect all real-time refinery data into a permanent data archive.
- The EIDI to be used as a separate data analysis platform from software systems located on the refinery control system network, so that plant operations are not compromised.
- The ability to define refinery asset in a hierarchical data model, so that similar asset types can be defined one time and replicated thereafter, which reduces deployment time and enables easier EIDI software maintenance.
- The ability to create time-sliced production events so that they are easily viewed and analyzed.

To create the EIDI digital plant database, the team recognized the need for good block flow and process flow diagrams to produce basic data structures and basic data navigation. A secondary requirement for creating the EIDI database is using process flow diagrams to check sensor locations in the refinery. These are used to configure the data definitions for physical elements. Process engineers can build performance calculations and create empirical models to infer laboratory results based on operating conditions.

A successful implementation of the EIDI at the South Texas refinery generates an opportunity to share results achieved with ProcIndustries management and other refineries. The company can now think about other possible insights from analysis of real-time data and events, such as the benefits of a common data visualization and analytics toolset and having subject matter experts use and share insights across different parts of the company.

References

Bascur, O.A. 1988. "A control data framework with distributed intelligence." *Advances in Instrumentation* 88:1553–1169.

Bascur, O.A. 1999. "Real-time process analysis to increase productivity." In *Proceedings of Second International Conference on Intelligent Processing and Manufacturing of Materials (IPMM 99)*, ed. J.A. Meech. Piscataway, NJ: Institute of Electrical and Electronics Engineers (IEEE).

Bascur, O.A. 2019. "Process control and operational intelligence in mineral and metallurgical processing." In *SME Mineral Processing & Extractive Metallurgy Handbook*, vol. 1, ed. R.C. Dunne, S.K. Kawatra, and C.A. Young. Englewood, CO: Society for Mining, Metallurgy & Exploration.

Bascur, O.A., and Halhead, M. 2013. "Energy effectiveness and sustainability management at Anglo American Platinum." In *Proceedings Copper 2013 International Copper Conference*, vol. 1, ed. C. Moscoso, J. Rosales, and A. Vio. Santiago: Chilean Institute of Mining Engineers (IIMCH). pp. 415–423.

Bascur, O.A., and Soudek, A. 2019. "Grinding and flotation optimization using operational intelligence." *Mining, Metallurgy & Exploration* 36(1):139–149.

Kennedy, J.P. 2019. "Retrospective: Then and now—looking toward the future." *Hydrocarbon Processing* (January). www.hydrocarbonprocessing.com/magazine/2019/january-2019/columns/retrospective-then-and-now-looking-toward-the-future.

Logue, C. 2019. "Integrating IT into process manufacturing." *Intech* (January/February). www.isa.org.

Steyn, J., Bascur, O.A., and Gorain, B. 2018. "Metallurgy analytics: Transforming plant data into actionable insights." *Mining Engineering* 70(9):18–29.

Turton, R., Bailie, R.C., Whiting, W.B., and Shaeiwitz, J.A. 1998. *Analysis, Synthesis, and Design of Chemical Processes*. Upper Saddle River, NJ: Prentice Hall.

Turton, R., Shaeiwitz, J.A., Bhattacharya, D., and Whiting, W.B. 2018. *Analysis, Synthesis, and Design of Chemical Processes*, 5th ed. Boston, MA: Prentice Hall.

Woods, D. 2005. *Successful Trouble Shooting for Process Engineers: A Complete Course in Case Studies*. Weinheim, Germany: Wiley-VCH.

Additional Reading

Bascur, O.A., Halhead, M., Garrigues, L., and Jarvis, M. 2016. "Mineral processing plant asset and energy optimization: The calming cloud over operations." In the *International Mineral Processing Congress (IMPC XXVIII) Proceedings*. Quebec City: IMPC.

Bascur, O.A., Hertler, C., and Wong, G. 2011. "Improving sustainability strategies in industrial complexes: Integration and collaboration." In *Proceedings of the EMC 2011*, ed. J. Harre. Lower Saxony, Germany: Clausthal-Zellerfeld.

Bascur, O.A., and Kennedy, J.P. 1995. "Measuring, managing and maximizing performance in petroleum refineries." *Journal of the Japanese Petroleum Institute*102.

Bascur, O.A., and Kennedy, J.P. 1996a. "Measuring, managing, maximizing refinery performance." *Hydrocarbon Processing* 75(1):111–116.

Bascur, O.A., and Kennedy, J.P. 1996b. "The industrial desktop—real-time business and process analysis to increase productivity in industrial plants." *Mining Engineering* (September).

Bascur, O.A., and Soudek, A. 2014. "Strategies for implementation of energy effectiveness and sustainability: Example Anglo American Platinum." *Presented at the 12th AusIMM Mill Operators' Conference: Achieving More with Less*, Australasian Institute of Mining and Metallurgy (AusIMM), Victoria, Australia.

Bascur, O.A., Vogus, C.B., and Bosler, W.H. 1992. "Long-term knowledge integration with OSHA PSM." In *NPRA Computer Conference Proceedings*, November 16–18. Washington, DC: National Petroleum Refiners Association.

Brynjolfsson, E., and McAFee, A. 2014. *The Second Machine Age: Work, Progress and Prosperity in a Time of Brilliant Technologies*. New York: W.W. Norton.

Duggan, K.J. 2012. *Design for Operational Excellence: A Breakthrough Strategy for Business Growth*. New York: McGraw-Hill.

Ferrari, A., and Russo, M. 2016. *Introducing Microsoft Power BI*. Redmond, WA: Microsoft Press.

Fogler, F.S., and LeBlanc, S. 1995. *Strategies for Creative Problem Solving*. Upper Saddle River, NJ: Prentice Hall.

Goldratt, E.M. 2014. *The Goal: A Process of Ongoing Improvement*, 4th rev. ed. Great Barrington, MA: North River Press.

Harari, Y.N. 2016. *Homo Deus: A Brief History of Tomorrow*. New York: Penguin.

Kennedy, J.P. 1994. Information Networks for Manufacturing Execution Systems. ISA Internal Course, Houston, TX.

Kennedy, J.P., and Bascur, O.A. 2002. "Influence of computer and information technology on process operations and business processes—a case study." In *Chemical Process Control VI: Assessment and New Directions for Research: Proceedings of the Sixth International Conference on Chemical Process Control*, ed. J.B. Rawlings, B.A. Ogunnaike, and J.W. Eaton. CACHE Series, vol. 98. New York: American Institute of Chemical Engineers (AICHE). pp. 7–11.

Kresta, J., MacGregor, J., and Marlin, T. 1991. "Multivariate statistical monitoring of process operating performance." *Canadian Journal of Chemical Engineering* C9:35–47.

Markman, A. 2012. *Smart Thinking: Three Essential Keys to Solve Problems, Innovate, and Get Things Done*. New York: Penguin.

Woods, D. 1994. *Problem Based Learning: How to Gain Most from PBL*. Hamilton, ON: W.L. Griffen Printing.

4
Advanced Analysis Using Unit Data and Event Templates

I believe you have to be willing to be misunderstood if you're going to innovate.

Jeff Bezos

Chapter Overview

In this chapter, we follow the refinery digital transformation team as they find more innovative ways to use the enterprise industrial data infrastructure (EIDI) information for continuous operational improvement. This chapter ties together many of the concepts described in earlier chapters, such as event frames and offline analysis, as the team attempts to understand and resolve a gap between scheduled production targets and actual production output.

They design and prototype a simple template, known as the *digital plant template*, as a basis for tracking the mysterious variance. The template components analyzed are simply production rate and consumable variables through various stages of the refining process. The team generates EIDI event frames to track for these operational stages, which when reviewed, determine exactly when and where these hidden production and energy losses occur. After pilot testing, they now have a reusable template that they can deploy to the entire refinery on a unit-by-unit basis. This chapter outlines four key steps the team deploys to solve this problem, significantly accelerating EIDI time to value.

This chapter describes a more advanced solution to analyze trouble time, idle time, and downtime, through the use of event frames and the digital plant template. This chapter will likely appeal to readers who are experienced, proficient EIDI users. If you are not quite in that category yet, we hope you will take away some ideas on how to better analyze a refinery or plant process through the tools mentioned here.

This chapter explains and sequences through most of the techniques required for this analysis. At the end of the chapter, there is a link to a manual that describes these techniques in more detail.

Innovative Use of EIDI Capabilities

Peter Argus' digital transformation team met to review the enterprise industrial data infrastructure (EIDI) deployment progress. Other than some small hiccups usually associated with installing a system of this scope, the initial implementation was going well.

After the initial use assessment part of the discussion concluded, Tom Jordan, the refinery plant manager, commented that the ProcIndustries South Texas refinery financial systems do not have sufficiently granular information about key operations production and consumables data. Much of this data is stored in the EIDI. Events that occur during the production process may shed light on causes of production losses. These events could be equipment startups or shutdowns, production set-up times, process idle times or downtimes, and unscheduled equipment shutdowns. Tom commented, "If we had a way to capture these events with relevant specifics, we could improve the overall flow of material in our refinery" (Bascur and Kennedy 2001; Bascur 2019; Bascur and Soudek 2019).

Tom told the digital transformation team that he wanted to capture these events in an effort to quantify the operating expenditures (OPEX). "This will eliminate what we have been trying to do for years in spreadsheets. If the EIDI can do it automatically, the need for individual spreadsheets will disappear and the information will be available to everyone," he stated. He then presented a chart showing typical ProcIndustries operational costs (Figure 4.1).

Tom explained that fuel is the largest operational cost followed by maintenance costs. The transition to a condition-based maintenance schedule is

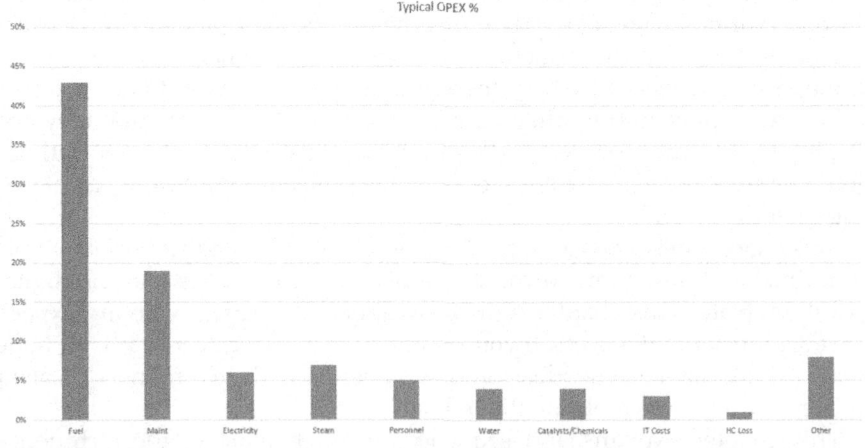

FIGURE 4.1
ProcIndustries refinery operational costs by percentage (HC loss = hydrocarbons lost during the refining process).

Advanced Analysis Using Unit Data and Event Templates

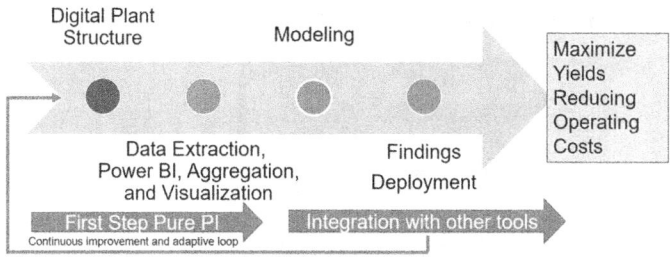

FIGURE 4.2
Four steps toward maximizing yields and reducing operating costs. (Courtesy of O.A. Bascur, OSIsoft LLC.)

helping reduce costs. EIDI-related data could potentially provide insights into other hidden energy costs.

Monica Armstrong (planning and economics coordinator), who helps define the daily production schedule, explained the current process to define refinery targets and key performance indicators (KPIs). "The team needs to show management how they can use the EIDI system to apply new methods of analysis that can identify process improvement opportunities—specifically, tools that analyze data during specific operating modes of plant equipment and calculate comparisons between planned targets and actual production yields."

Peter and Monica realized that they need to have operational events classified in real time and quickly calculate production losses. By possessing this real-time data and the production loss times, they can improve on the parameters used as input to the refinery linear programming (LP), used for scheduling and optimization.

The transformation team constructs a four-step process to show management the flow of how the EIDI data is used to analyze and solve problems, as shown in Figure 4.2.

Step 1: Develop Unit Process Templates Using Plant Block Diagrams

"Let's discuss how we build a digital plant model," said Peter. Turton et al. (2018) describes best practices for building process block and flow diagrams. Updated process block, flow, and piping and instrumentation diagrams, used with real-time data, achieve best practices for process monitoring (Bascur 2019).

Online, real-time configurable calculations can be created using EIDI analytics, with event frames created to monitor the exact state of the process and its duration This capability enables the team to use an EIDI template to generate operational modes for all units. These operational modes are then used to set the event framing start and end times, which generates the minimum, maximum, total, and average and standard deviation value for each variable in the event template. The ability to integrate data from multiple sources and

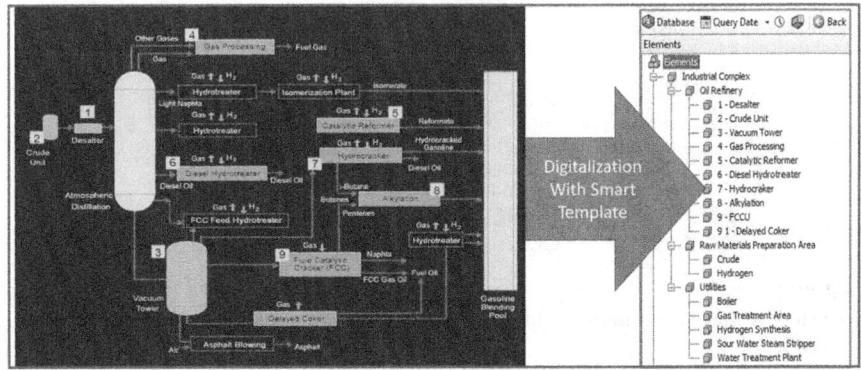

FIGURE 4.3
Oil refinery block diagram transformed to a digital plant schema. (Courtesy of O.A. Bascur, OSIsoft LLC.)

apply logic combining real time and external channels offers a variety of methods for determining KPIs.

Our journey starts by configuring unit templates with live process variables. Let's take a look at our refinery's process block diagram, which is shown in Figure 4.3.

Figure 4.3 shows the process block diagram in relation to the hierarchical structures represented by a *single, reusable unit template*. Peter explained that the unit template is part of the EIDI and the concept has been used in other process industries for many years.

The transition from calendar- to condition-based maintenance provides the ability to capture information about various states of equipment. For example, determining the number of starts and stops over time can help to determine appropriate maintenance cycles. The ability to capture these events along with related information such as leading indicators can provide valuable insight into the prevention of future failure.

Each time a unit changes its operational state, it records an event frame. The event frame subsystem marks the time intervals for further processing and analysis. The time intervals determine aggregation of actual production and consumable data values, to estimate production and consumption losses (totals, averages, standard deviation, minimum and maximum).

The EIDI contains many calculations and algorithms to handle time-series data, such as interpolation of sampled laboratory data with real-time production data to group them together when developing process models (Steyn et al. 2018; Bascur and Soudek 2019).

The EIDI unit template event frame captures process unit rates and consumables to automatically calculate average production rate, total energy, and water and air consumed for each operating mode. Table 4.1 shows the typical results.

TABLE 4.1
Event Frame Data Created for Extraction and Analysis

Event Frame	Asset	Start	End	Duration	Mode	Process Feed Rate	Electricity Consumption	Water Consumption	Other Variables
Analysis template 20120725	Boiler	August 1, 2019 12:44:00 p.m.	August 2, 2019 3:55:00 a.m.	15:01:00	Running	131.5	84.4	20.5	XX.X
Trouble state duration	Fluid catalytic cracking unit (FCCU)	August 2, 2019 3:55:00 a.m.	August 2, 2019 4:00:00	00:05:00	Trouble	0.0	30.0	10	X.XX
Down state time	Desalter	August 2, 2019	August 2, 2019	00:12:00	Down	0.0	15.0	5.0	X.XX
Running OK state duration	Alkylation	August 2, 2019	August 2, 2019	03:46:00	Running	95.0	78.0	45.0	X.XX
Maintenance state duration	Vacuum tower	August 2, 2019	August 2, 2019	00:10:00	In maintenance	0.0	20.0	5.0	X.XX

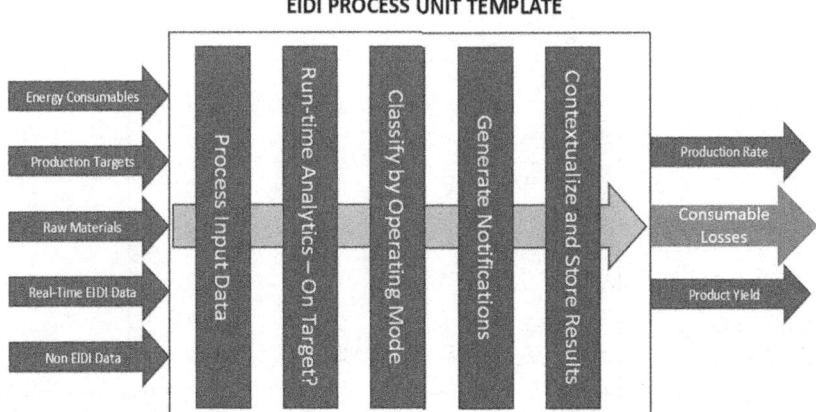

FIGURE 4.4
Smart process unit template components.

Monica explained further, "We can derive the operational status by monitoring real-time production targets and comparing actual results. Each unit calculation is identical, creating event frames to indicate various operational modes." Figure 4.4 shows the basic building blocks of a process unit template, which can be replicated for other units and refineries.

Monica offered, "The same production unit template can be reused at other ProcIndustries refineries to help them get started when implementing their EIDI. It's best to start simple and not reinvent the wheel."

Figure 4.5 shows ProcIndustries' production targets and process variables, which are the inputs to the process unit template. Monica pointed out the system side of real-time intelligence that uses powerful analytics and event-framing capabilities: "Are we on target?" "Are we satisfied?" represents the human side, providing information and operational insights for further decision-making (Bascur and Halhead 2013).

A unique data model is the skeleton for all areas of the process plant. This data model is like DNA; it is an object model that has the key attributes that describe the performance of a process unit. In this case, a process unit takes a process feed rate with reagents, energy, fuel, water, and air to generate product stream(s).

Monica explained that they use production rate, consumable electricity, and water and air totals in the unit template and five operating state definitions: "running OK," "trouble," "idle," "down," or "in maintenance." During production runs, this modular, reusable unit template calculates and segments time into one of those operational modes.

Advanced Analysis Using Unit Data and Event Templates 93

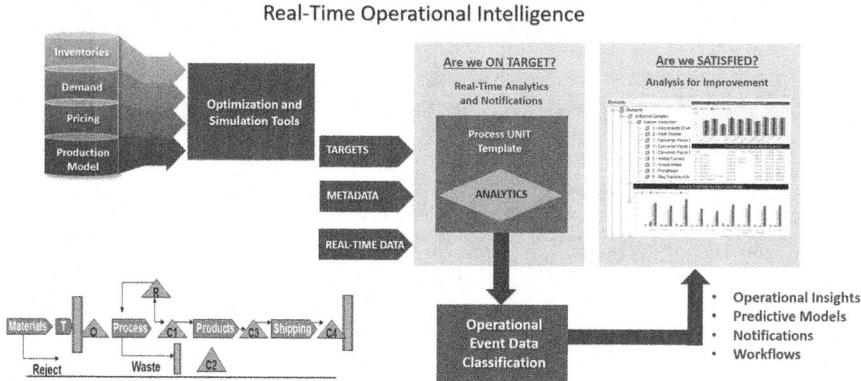

FIGURE 4.5
Smart unit template to evaluate overall refinery production performance. (Courtesy of O.A. Bascur, OSIsoft LLC.)

Step 2: Analyzing and Visualizing Operational Variance from the Generated Events

Once the unit template is applied to each of the refinery unit elements, the respective attributes are configured with real-time data stream tags. As shown in Figure 4.6, the EIDI human interface software (OSIsoft's PI Vision)

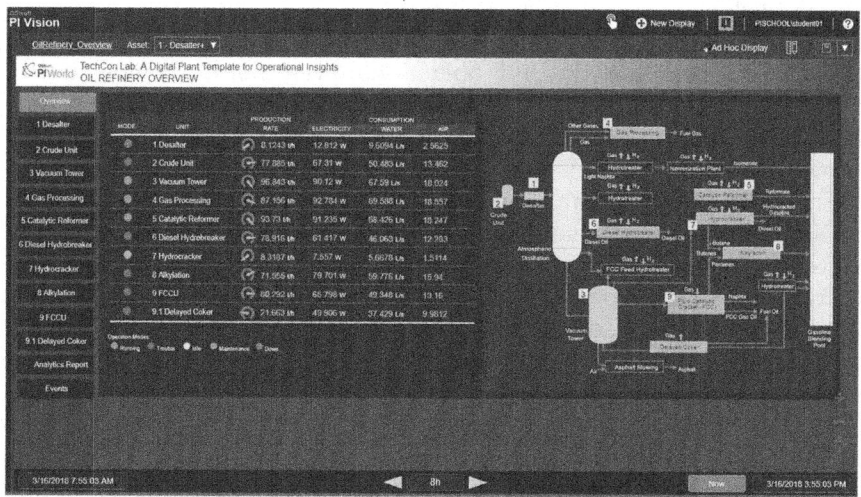

FIGURE 4.6
Refinery overview dashboard. (Courtesy of O.A. Bascur, OSIsoft LLC.)

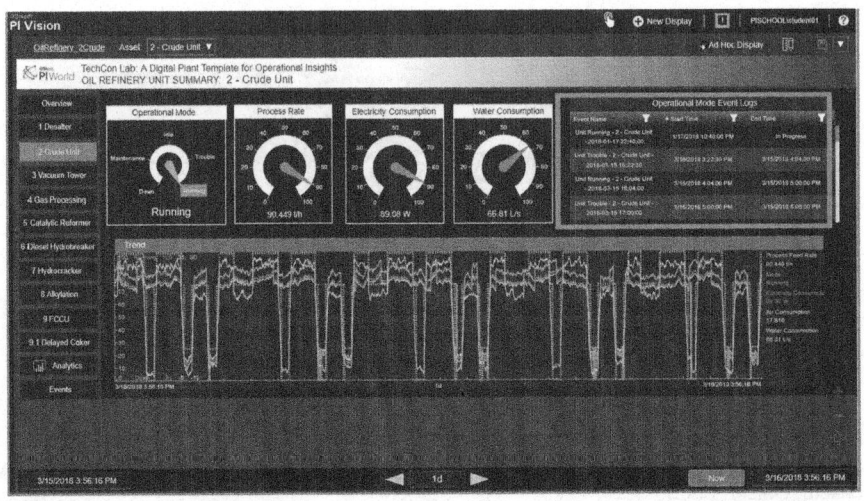

FIGURE 4.7
Crude unit relative dashboard with unit template attributes and operating modes. (Courtesy of O.A. Bascur, OSIsoft LLC.)

displays the real-time information for all refinery units, the unit template operating modes, as well as production and consumable data.

To show a more detailed KPI dashboard for that specific unit, the user can simply click on the appropriate button (Figure 4.7).

Figure 4.7 shows a real-time crude unit dashboard displaying the key process unit attributes and the operational event notifications (shown in the *rectangle* at the top right) generated by the EIDI event frame subsystem. By clicking one of these notifications, a second display is generated, as shown in Figure 4.8.

The power of event framing is the ability to compare similar events, operating modes, and batches/lots against each other through online data visualization tools such as PI Vision, or to export these events and their corresponding data to other software systems for offline analysis. As previously stated, users can compare an ideal batch/event/lot data profile against other similar production runs to determine what exactly causes an ideal run.

Figure 4.8 shows the production and consumable losses, such as electricity and water, for selected operational time events.

Monica went on, "By analyzing the different operational states, we can see the variance between an expected and an actual result. The expected results are those specified in the budget or in current production schedules, produced by my planning and economics team. We get the production schedule targets from a LP model that incorporates the inventories,

Advanced Analysis Using Unit Data and Event Templates 95

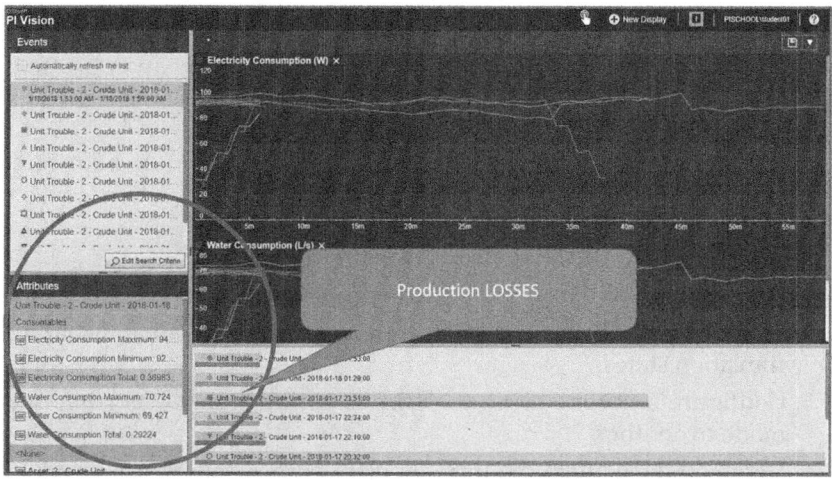

FIGURE 4.8
Event frame display showing crude unit production and consumable losses. (Courtesy of O.A. Bascur, OSIsoft LLC.)

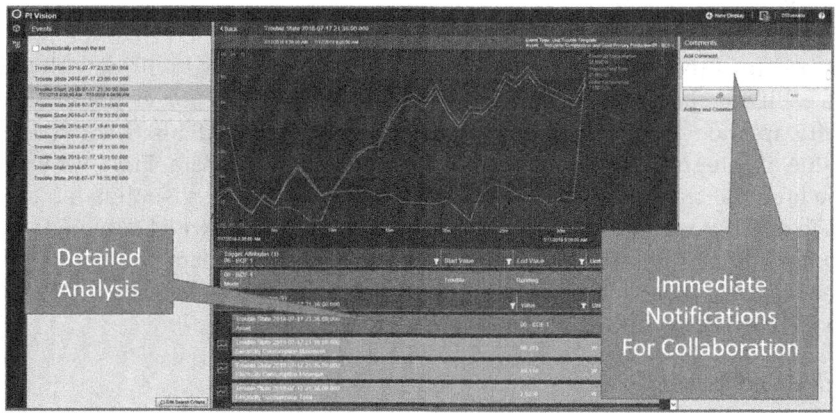

FIGURE 4.9
Unit analysis of operational variables with real-time notification ability. (Courtesy of O.A. Bascur, OSIsoft LLC.)

customer orders, and process unit parameters for several types of raw materials and processing strategies."

Figure 4.9 shows the detailed information for a selected process unit, with event start and end times, attribute history, and when the event was active.

Classifying Asset Behavior for Process Improvement

Peter surmised: "This frees up valuable time for engineers and operators to make quicker decisions. The EIDI PI Vision displays present the right data to solve most of our difficult operational problems."

The digital transformation team then put the following ideas into practice:

1. Build a digital plant template for each process unit.
2. Identify input data streams for each unit template.
3. Define EIDI event frames for each unit production mode.
4. Identify conditions to indicate various modes of operation including transition states.
5. Configure real-time asset analytics to record transitions from one mode to another.
6. Define and calculate the unit template output values (e.g., production rate, yield).
7. Configure notifications to alert people and systems of process mode changes.

Step 3: Employing Offline Visualization Tools Using Unit Process Template Data

The business value of the data dramatically increases each time the data is reused for other business and operational use cases. We will now discuss how EIDI can supply contextualized, high-fidelity data for visual analysis, software modeling, and advanced analytics tools. We will also introduce the value of integrating non-EIDI analytics with real-time data. First, we examine how the team analyzes EIDI data through advanced visualization tools, such as Microsoft Power BI or Tableau. Our ability to extract EIDI data in a contextualized format allows for offline analysis. The data can be used for data cleansing or predictive analytics to apply machine learning (ML) using tools like Microsoft ML Studio or Python and R.

Advanced Visual Analytics

An effective real-time data infrastructure enables automatic, configurable import of event-framed operational data for the entire refinery. Open database connectivity (ODBC) or published data sets provide a basis for these analyses.

Figure 4.10 shows the Microsoft Power BI desktop application. Selecting the "Get Data" ribbon control (*left circle*) updates the EIDI event frame data. Using Microsoft Power BI and the Azure Cloud, the team can schedule an automatic update and publish the information. The ProcIndustries team decides to use an EIDI ODBC query for a quick start.

Analytics tools allow multidimensional visualization of the extracted production and operations data (Ferrari and Russo 2016). The team can generate alternate views of the data displays, trends, and reports.

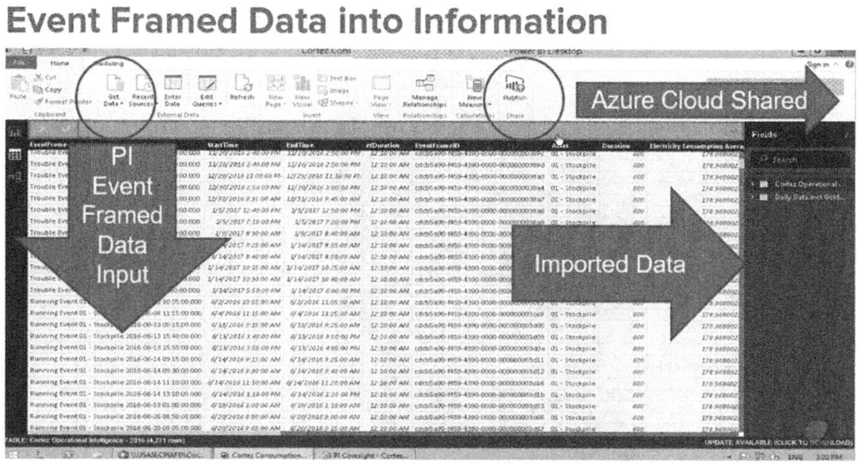

FIGURE 4.10
Microsoft Power BI desktop ingesting contextualized event-based data. (Courtesy of O.A. Bascur, OSIsoft LLC.)

Figure 4.11 shows a multidimensional data cube containing time interval data for the units, material type, operating modes, and aggregated data. Additional data for operating modes, shifts, crews, and raw material feed types allows for additional analyses. The results are cleansed before use in process modeling to highlight hidden losses by operating mode.

Tools such as Microsoft Power BI allow the user to publish dashboard reports using Azure Power BI (see Figure 4.10, "Publish" shown in the *circle on the top right of the figure*).

Figure 4.12 shows a Microsoft Power BI dashboard displaying water consumption for all process units for all operational states. Monica pointed out, "This lets us visualize the hidden losses of production, water, utilities, and any consumption variables chosen in the unit template and event frame template." These self-service tools allow everyone to see how well the refinery is operating. Viewing the data in many forms enables people to take corrective actions for continuous improvements and for developing predictive analytic models.

"Power BI dashboards allow us to integrate the EIDI Asset Framework calculations with event frame data. We can utilize data pivots for specific times and events to produce detailed performance analyses." Peter explained, "By making these operational events visible for all process units, we can automatically calculate an overall production effectiveness index using Microsoft Power BI." The dashboard provides insights regarding operational modes in relation to time.

This strategy used to estimate the overall production effectiveness (OPE) is called "follow the money" (Plourde 2016, 2019; Plourde et al. 2017). It assists in improving the coordination of the supply chain in large industrial

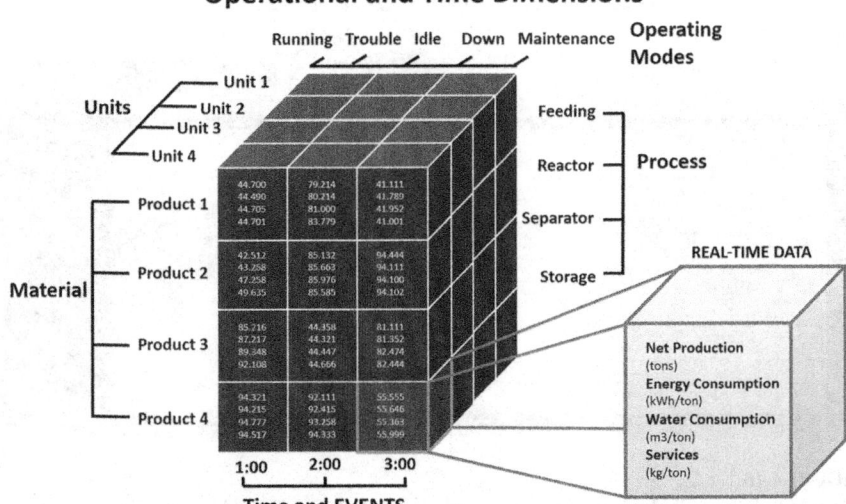

FIGURE 4.11
Event frame cube of data and event pivots for business intelligence analysis. (Courtesy of O.A. Bascur, OSIsoft LLC.)

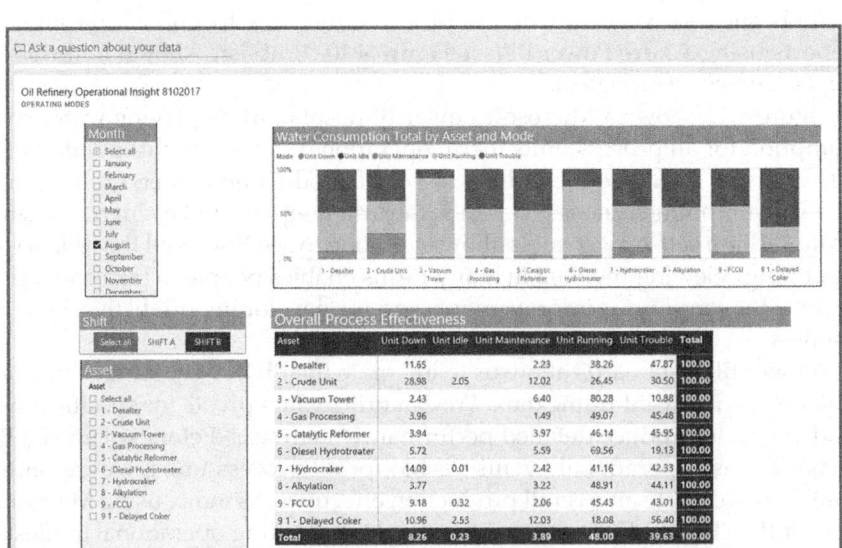

FIGURE 4.12
Microsoft Power BI dashboard showing overall production effectiveness. (Courtesy of O.A. Bascur, OSIsoft LLC.)

Advanced Analysis Using Unit Data and Event Templates 99

complexes. Bill Roberts, vice president of operations, has often expressed that the ability to detect and quantify these types of losses are enough to justify the implementation of the EIDI system.

The ability to combine data from different data sets using Power BI allows the team to access information from multiple data sets. For example, the dashboard can have tiles from refinery inventory and their terminals. The team learned that they should be creative in the design of the EIDI's asset object model and in event frame generation. That is, they should look to include any EIDI or non-EIDI data that will help them with their analysis.

Peter and some of the team members configured the EIDI Asset Framework in one day. By applying a straightforward unit data model approach and a standardized nomenclature to define assets in the refinery, they were able to evaluate production versus planning targets as requested by the refinery manager. Refinery personnel started creating ad hoc reports using Power BI.

Figure 4.13 shows a cell phone screenshot of the event-framed operational data analysis. "Once the operational event results are regularly updated

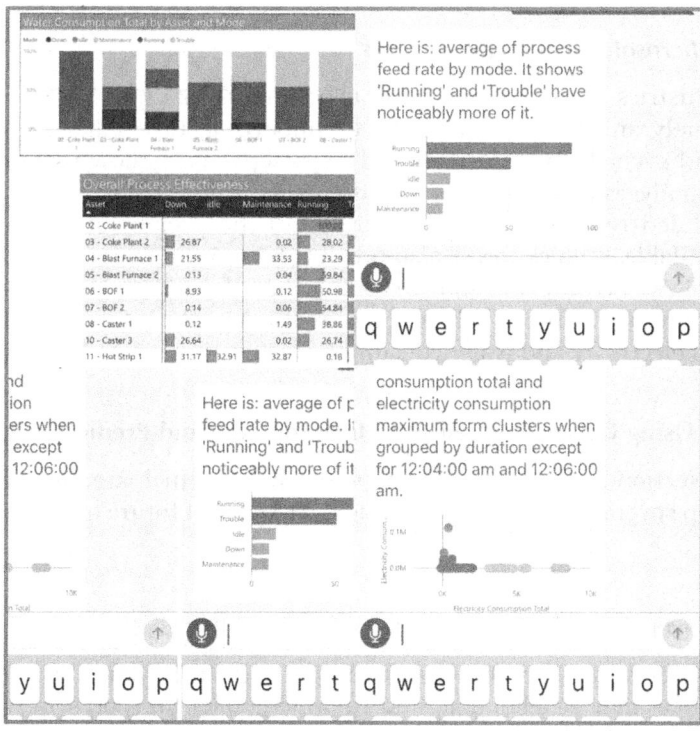

FIGURE 4.13
Microsoft Azure Power BI using Cortana artificial intelligence. (Courtesy of O.A. Bascur, OSIsoft LLC.)

using Power BI, you can use Cortana on your devices to provide suggestions for operational data analysis," said Peter.

The smart unit template with an operational variance algorithm assists companies by

- Characterizing the OPE;
- Reporting production and consumables losses when the plant is operating in trouble (invisible losses), as well as identifying down times, idle times, and losses; and
- Building predictive analytic models using running OK data sets.

"Now, that we have achieved a successful pilot, we need to convince others at ProcIndustries to move forward with our process improvements," said Peter.

During the team dinner afterward, they congratulated themselves on their achievement toward demonstrating the value of the EIDI, as the ProcIndustries CEO and Bill Roberts had asked. They also discussed how to prioritize the many opportunities detected by this strategy.

Using Microsoft Excel Analytics Tools

ProcIndustries personnel needed to extract EIDI data into Microsoft Excel. Fortunately, an add-in was available for importing current, historical, statistical, and event frame information. The objective is to allow users to create ad hoc analyses on a variety of data types.

Scheduled reports, such as a daily production report at midnight, are now available. Data export is also possible for non-EIDI analytic tools, such as MATLAB®. Figure 4.14 shows results when the equipment was running along with the predicted values and regression analysis for crude unit feed rate and electricity and water consumption.

Step 4: Using Contextualized Data for Modeling and Predictive Analysis

In this section, we delve into using soft sensors, which are derived values based on several different inputs and how to predict future outcomes, based on ML.

Data-Driven Analytics

Soft sensor values are the combination of data from multiple sources. The sources may be laboratory and equipment data, and soft sensors can be predictors of both laboratory and equipment data. To develop a soft sensor, the team combined the "running OK" data set with laboratory operational

Advanced Analysis Using Unit Data and Event Templates

PI Datalink data extraction when the process area is Running OK. And linked the TIMES that the other variables are running OK							
Start Time	*-7d			Filter Expression			
End Time	*			('\\PISRV01\REF.2 - Crude Unit.Mode'="Running")			
Holy Tag	\\PISRV01\REF.2 - Crude Unit.Mode			Electricity = Intercept + Coef1 * Water + Coef2 * Feed Rate			
When the unit is happy we can develop a predictive model							
Here put the result		Y Electricity	X1 Water	X2 Process Feed Rate			
PI CompDat		PITimeDat					
		\\PISRV01\REF.2 - Crude Unit.Electricity Consumption	\\PISRV01\REF.2 - Crude Unit.Water Consumption	\\PISRV01\REF.2 - Crude Unit.Process Feed Rate	Prediction	MultiLinear Regression using Excel DataPack	
Number of Values: 102							
12-Jan-18 17:20:00 Running		91.13020557	68.34765418	87.88004579	91.128302	SUMMARY OUTPUT	
12-Jan-18 19:37:00 Running		85.75056985	64.31292739	88.15976077	85.748932		
12-Jan-18 20:50:00 Running		97.15239763	72.86429822	88.47486878	97.149693	Regression Statistics	
12-Jan-18 22:16:30 Running		90.86499989	68.14874992	87.53406421	90.863309	Multiple R	0.9999486
13-Jan-18 00:37:00 Running		85.85541526	64.39156145	89.38395624	85.853089	R Square	0.9998973
13-Jan-18 01:51:00 Running		91.3533711	68.51502832	88.84825089	91.350913	Adjusted R S	0.9998952
13-Jan-18 03:20:00 Running		93.51325381	70.13494036	87.11578197	93.511588	Standard Err	0.0495176
13-Jan-18 05:38:30 Running		86.25379591	64.69034694	91.53183257	86.250246	Observation	102
13-Jan-18 06:57:00 Running		81.77426567	61.33069925	87.61537813	81.773241		
13-Jan-18 08:26:00 Running		88.82164873	66.61623655	87.64041579	88.820059	ANOVA	
13-Jan-18 10:49:00 Running		89.08079993	66.81059994	87.0187545	89.079535		df SS MS
13-Jan-18 11:53:00 Running		88.63104557	66.47328418	90.09318009	88.628109	Regression	2 2362.903 1181.452
13-Jan-18 13:20:30 Running		86.95842838	65.21882129	87.66940377	86.956968	Residual	99 0.242747 0.002452
13-Jan-18 15:43:00 Running		83.2804411	62.46033082	88.01267761	83.279078	Total	101 2363.146
13-Jan-18 16:54:00 Running		88.56327784	66.42245838	91.66814874	88.559472		
13-Jan-18 18:20:00 Running		87.68218972	65.76164229	93.62482914	87.677366	Coefficients andard Err t Stat	
13-Jan-18 21:01:00 Running		85.3755043	64.03162822	92.35438224	85.371566	Intercept	0.0540231 0.194969 0.277085
13-Jan-18 21:55:00 Running		83.32917042	62.49687782	87.10884477	83.328305	X Variable 1	1.3332291 0.001369 974.1491
13-Jan-18 23:21:00 Running		95.59803973	71.6985298	89.61958424	95.594821	X Variable 2	-0.0005553 0.002045 -0.27149

FIGURE 4.14
Importing operations data to Excel filtered by status (in "running OK" mode). (Courtesy of O.A. Bascur, OSIsoft LLC.)

data to create a predictive model. They then incorporated the model into online asset-based template calculations. Using soft sensors increases the value of the control strategies, avoiding process constraints and moving into optimal operational states.

The operational variables are classified into (1) manipulated, (2) controlled, and (3) disturbances. "Selecting these variables lets us predict the results of the target variables (target tags). This is done based on the effect of the disturbances, which suggests how to move the manipulated variables to achieve the desired objectives. The target variable is what you model based on the operating variables. If it is a soft sensor, it can be a controlled variable," said Peter.

Having all process units available simplifies the selection of the manipulated variables to avoid constraints. This way, optimal yield can be achieved while reducing operational costs.

By extracting subsets of the "running OK" data into Excel spreadsheets, the team creates tabular data, which is easily imported into ML tools. Referring to the spreadsheet shown in Figure 4.14, it shows the algorithm that interpolates the requested data for further analysis with the ML tools. The ability to integrate contextualized real-time and event-based information and then pass the results to external modeling tools is an effective form of data-driven analytics.

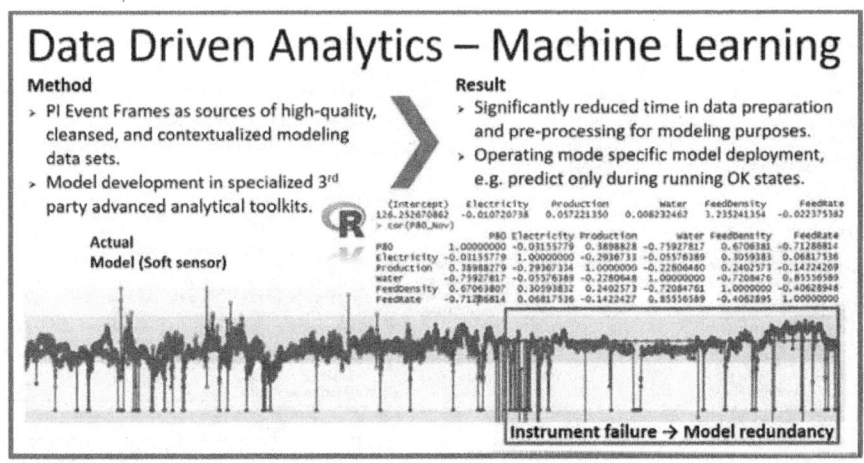

FIGURE 4.15
Predictive model showing a soft sensor generated from the running OK mode data. (Courtesy of O. A. Bascur, OSIsoft LLC.)

Figure 4.15 shows the Microsoft Azure Machine Learning Studio objects, which are used to generate a predictive analytics model. Predictive analytics is the art of building and using models that make predictions based on patterns extracted from historical data. An example is sample laboratory data, such as percent assays, turbidity, temperature, and pressure distillation profiles.

Figure 4.15 shows the soft sensor results using EIDI analytics to calculate a predictive model. This model is generated with a dataset in Microsoft's Machine Learning Studio to calculate the coefficients for a multilinear regression algorithm. The real-time model shows the predictive soft sensor data for a particle size analyzer (Steyn et al. 2018). This particle size analyzer soft sensor is used in implementing an advanced control strategy along with the real-time sensor. It acts as a redundant sensor with the physical one, if the physical sensor is not working correctly, as shown in Figure 4.15.

The ability of time- or length-based data aggregation within the data infrastructure provides the necessary augmented data for analysis such as averages, standard deviations, totals, modes, as well as minimum and maximum for the many operating states. It has been found that by having the data subset contextualized for particular operational states, the development of predictive models is simplified. It is easy to understand that if everything is in the "running OK" mode, the models will be completely different than if there is a disturbance such as an equipment failure or a fire.

For further information on how to use Microsoft's Machine Learning Studio and other tools with these unit event templates, see the "Additional Reading and Template Implementation Materials" box at the end of this chapter.

What You Should Take Away

Using an EIDI for continuous improvement and innovation strategies represents the preferred way of doing business. The ProcIndustries digital transformation team demonstrated that it is critical to view tasks as steps in their process instead of unrelated events. They used their EIDI real-time data infrastructure to analyze operating information by selecting data for specific assets at specific points in time, filtered by operating mode. Through this, they can learn how to identify problems and strengthen their capabilities to address and prevent them.

They designed and prototyped a simple template—the digital plant template—as a basis for tracking variances through various stages of the refining process that had not been easy to find. EIDI event frames helped to identify hidden production and energy losses.

The team used five key steps to solve this problem:

1. They developed unit process templates using plant block diagrams.
2. They defined and configured EIDI event frames that were triggered by determining which of the preconfigured modes the process was in at any one time.
3. They analyzed and visualized the collected data by utilizing EIDI online visualization tools. The team was able to view data in several formats: dashboards, historical trend data, and tabular data in Excel spreadsheets. They were able to filter data by operational mode for all these data presentation formats to help determine what caused hidden losses and variances from expected results.
4. They extracted contextualized EIDI data for offline visual analysis using tools like Microsoft Power BI pivot tables. The data was filtered by process stage and operating mode so that the team could filter and view what occurred when the process was in a specific operating mode such as "running OK" or "idle."
5. They reused the extracted EIDI data for modeling and by artificial intelligence (AI) predictive analytic tools, such as MATLAB, Python, R, and ML.

In a continuous-improvement program, there is a laser focus on variances: where and why they happen, and what to do about them. At the ProcIndustries refinery, the production schedule is derived from a linear programming model that contains all process unit parameters for several types of products.

In their EIDI data analysis work, Peter collaborated with other members of the South Texas refinery team to transform operations and equipment data into time-series contextualized information about all refinery process units. The team used Microsoft Power BI, a tool that enables them to analyze the EIDI data in depth to uncover what was causing these operating variances, limiting the refinery's effectiveness.

They came to understand that framing events through the EIDI system and generating notifications to indicate when a process unit diverged from production targets was highly effective in managing refinery abnormalities. By focusing on these events, the team became proactive in addressing issues before they become serious problems. This reduced losses from underperforming or inefficient equipment and processes.

The team incorporated its collective knowledge into EIDI asset templates. In turn, these asset templates help to detect important events during operations. The ability to associate real-time data in the asset data model helps to asset current operations. This is one of the most valuable activities a team can do, as it seeks to optimize performance in an industrial process plant.

By using the EIDI, the ProcIndustries refinery staff can record the resources consumed in each operating unit. These consumables are easily aggregated by total cycle time or calculated according to resources ingested by every subset of events during the cycle. This capability provides a cost approximation for each activity, which can be measured against the value it creates for the final products.

The team recognizes that improvements in the South Texas refinery will have far-reaching ramifications. The successful implementation of the EIDI at the South Texas refinery provides a basis throughout ProcIndustries. Other refineries and executives at headquarters will benefit from the lessons learned. The EIDI simplifies disparate systems at each location. The enterprise will be able to

- Collaborate on real-time data and event problem solving;
- Share a common data visualization and analytics tool set;
- Establish a common platform for notifications;
- Leverage insights of subject matter experts across different parts of the enterprise; and
- Deploy remote monitoring and diagnostic centers, allowing refinery personnel to focus on production.

ADDITIONAL READING AND TEMPLATE IMPLEMENTATION MATERIALS

For more detailed information on using the digital plant template, with step-by-step implementation instructions, refer to www.osisoft.com/digitalplant.

About a third of the way down that page, you will see PDF attachments that will aid you in deploying the components and the analysis sequence described in this chapter.

References

Bascur, O.A. 2019. "Process control and operational intelligence in mineral and metallurgical processing." In *SME Mineral Processing & Extractive Metallurgy Handbook*, vol. 1, ed. R.C. Dunne, S.K. Kawatra, and C.A. Young. Englewood, CO: Society for Mining, Metallurgy & Exploration. pp. 277–316.

Bascur, O.A., and Halhead, M. 2013. "Energy effectiveness and sustainability management at Anglo American Platinum." In *Proceedings Copper 2013 International Copper Conference*, vol. 1, ed. C. Moscoso, J. Rosales, and A. Vio. Santiago: Chilean Institute of Mining Engineers (IIMCH). pp. 415–423.

Bascur, O.A., and Kennedy, J.P. 2001. "Real-time information management for asset optimization." In *Mineral Processing Plant Design, Practice, and Control*, ed. A.L. Mular, D.N. Halbe, and D.J. Barratt. Littleton, Co: Society for Mining, Metallurgy & Exploration.

Bascur, O.A., and Soudek, A. 2019. "Grinding and flotation optimization using operational intelligence." *Mining, Metallurgy & Exploration* 36(1):139–149.

Ferrari, A., and Russo, M. 2016. *Introducing Microsoft Power BI*. Redmond, WA: Microsoft Press.

Plourde, M. 2016. "Heavy haul equipment logistic strategies in iron ore mining." Presented at the 2016 Users Conference, San Francisco, Process Industries, Transportation, and Supply Chain. www.osisoft.com/Presentations/Heavy-Haul-Equipment-Logistic-Strategies-in-Iron-Ore-Mining.

Plourde, M. 2019. "PI System at the heart of a mining integrated operations center (ArcelorMittal)." Presented at the 2019 PI World, San Francisco, Mining, Materials, and Supply Chain. www.osisoft.com/Presentations/The-PI-system-at-the-heart-of-a-Mining-Integrated-Operations-Center--ArcelorMittalx/.

Plourde, M., Bascur, O.A., Paquet, S., and Gervais, D. 2017. Digital innovation in modern engineering and operational excellence. Presented at the 2017 SME Annual Conference and Expo, Denver, February 19–22.

Steyn, J., Bascur, O.A., and Gorain, B. 2018. "Metallurgy analytics: Transforming plant data into actionable insights." *Mining Engineering* 70(9):18–29.

Turton, R., Shaeiwitz, J.A., Bhattacharya, D., and Whiting, W.B. 2018. *Analysis, Synthesis, and Design of Chemical Processes*, 5th ed. Boston, MA: Prentice Hall.

Additional Reading

Bascur, O.A. 1988. "A control data framework with distributed intelligence." *Advances in Instrumentation* 88:1553–1169.

Bascur, O.A. 1999. "Real time process analysis to increase productivity." In *Proceedings of Second International Conference on Intelligent Processing and Manufacturing of Materials (IPMM 99)*, ed. J.A. Meech. Piscataway, NJ: Institute of Electrical and Electronics Engineers (IEEE).

Bascur, O.A., Hertler, C., and Wong, G. 2011. "Improving sustainability strategies in industrial complexes: Integration and collaboration." In *Proceedings of the EMC 2011*, ed. J. Harre. Lower Saxony, Germany: Clausthal-Zellerfeld.

Bascur, O.A., and Kennedy, J.P. 1995. "Measuring, managing and maximizing performance in petroleum refineries." *Journal of the Japanese Petroleum Institute* 102.

Bascur, O.A., and Kennedy, J.P. 1996a. "Measuring, managing, maximizing refinery performance." *Hydrocarbon Processing* 75(1):111–116.

Bascur, O.A., and Kennedy, J.P. 1996b. "The industrial desktop—Real time business and process analysis to increase productivity in industrial plants." *Mining Engineering* (September).

Bascur, O.A., and O'Rourke, J. 2019. "Measuring, managing and transforming data for operational insights." In *Smart Manufacturing*, ed. M. Soroush, M. Baldea, and T. Edgar. New York: American Institute of Chemical Engineers.

Bascur, O.A., and Soudek, A. 2014. "Strategies for implementation of energy effectiveness and sustainability: Example Anglo American Platinum." Presented at the 12th AusIMM Mill Operators' Conference: Achieving More with Less, Australasian Institute of Mining and Metallurgy (AusIMM), Victoria, Australia.

Bascur, O.A., and Soudek, A. 2019. "Grinding and flotation optimization using operational intelligence." *Mining, Metallurgy & Exploration* 36(1):139–149.

Bascur, O.A., Vogus, C.B., and Bosler, W.H. 1992. "Long-term knowledge integration with OSHA PSM." In *NPRA Computer Conference Proceedings*, November 16–18. Washington, DC: National Petroleum Refiners Association.

Bodington, C.E., and Shobrys, D.F. 1999. "Optimize the chain supply." In *Advanced Process Control and Information Systems for the Process Industries*, ed. L.A. Kane. Houston: Gulf. pp. 236–240.

Brynjolfsson, E., and McAfee, A. 2014. *The Second Machine Age: Work, Progress and Prosperity in a Time of Brilliant Technologies*. New York: W.W. Norton.

Duggan, K.J. 2012. *Design for Operational Excellence: A Breakthrough Strategy for Business Growth*. New York: McGraw-Hill.

Faivre, B., and Alexander, T. 2016. "Deschutes: Better data for better beer." OSIsoft Case Studies and Testimonials. https://www.osisoft.com/pi-system/case-studies-and-testimonials/all-case-studies/Deschutes2016brief/.

Fogler, F.S., and LeBlanc, S. 1995. *Strategies for Creative Problem Solving*. Upper Saddle River, NJ: Prentice Hall.

Goldratt, E.M. 2014. *The Goal: A Process of Ongoing Improvement*, 4th rev. ed. Great Barrington, MA: North River Press.

Harari, Y.N. 2016. *Homo Deus: A Brief History of Tomorrow*. New York: Penguin.

Kennedy, J.P. 1994. *Information Networks for Manufacturing Execution Systems*. Houston, TX: ISA Internal Course.

Kennedy, J.P., and Bascur, O.A. 2002. "Influence of computer and information technology on process operations and business processes—A case study." In *Chemical Process Control VI: Assessment and New Directions for Research: Proceedings of the Sixth International Conference on Chemical Process Control*, ed. J.B. Rawlings, B.A. Ogunnaike, and J.W. Eaton. CACHE Series, vol. 98. New York: American Institute of Chemical Engineers (AICHE). pp. 7–11.

Kresta, J., MacGregor, J., and Marlin, T. 1991. "Multivariate statistical monitoring of process operating performance." *Canadian Journal of Chemical Engineering* C9:35–47.

Markman, A. 2012. *Smart Thinking: Three Essential Keys to Solve Problems, Innovate, and Get Things Done*. New York: Penguin.

Woods, D. 1994. *Problem Based Learning: How to Gain Most from PBL*. Hamilton, ON: W.L. Griffen Printing.

Woods, D. 2005. *Successful Trouble Shooting for Process Engineers: A Complete Course in Case Studies*. Weinheim, Germany: Wiley-VCH.

5

The Humans behind the Data: Visualization and Collaboration

[Real-time process management] is by nature strategic: It creates possibilities that did not exist before, but without innovative processes, all you have accomplished is the automation of current bad processes and procedures, and sometimes this only does bad things faster and more accurately.

Pat Kennedy

Chapter Overview

In this chapter, we take a short hiatus from describing the technology behind the South Texas refinery enterprise industrial data infrastructure (EIDI) initiative. Instead, we discuss the human aspects of making such an initiative successful. Technology by itself will not solve business problems. Workers who are willing to collaborate with other teams to improve the business are essential for a successful outcome.

We describe some of the essential topics that the ProcIndustries management team needs to address and plan for to ensure that South Texas refinery employees buy in to using the new EIDI system productively. Company leadership also needs to effectively communicate impending changes to workflows and what is expected of the employees. Uncertainty should be reduced as much as possible during the initial phases of the deployment. Team managers should make sure that employees understand their new responsibilities and new workflows, motivating everyone to pull in the same direction for refinery operations improvement.

We also present ProcIndustries' plan to educate knowledge workers to effectively use the EIDI and make sure workers get the help they need during the transition. As the deployment progresses, people become self-sufficient so they can create their own customized data views.

The week after sending refinery data to the enterprise industrial data infrastructure (EIDI) and building some configuration templates, the team from the South Texas refinery regrouped. They had recently performed some successful pilot tests to make sure the EIDI was collecting complete and accurate data from the refinery's control system. They had also prototyped a digital plant template to help them analyze efficiency during various operating states. Now, it was time to address the human aspects of the continuous-improvement project. They needed to determine how to best present the new EIDI information to workers in the most beneficial way.

Because of the meeting importance, Bill Roberts, the vice president of operations and executive sponsor of the EIDI project, attended. Bill asked the team members to keep in mind the following five themes:

1. Present the EIDI data in a role-based manner, so that workers consume the information that is most important for them to be productive.
2. The team must make sure that all users stay informed regarding deployment plans and are regularly informed on significant progress or delays that occur. If there is a change of design, the team must make sure that the appropriate people receive the news, so there are no surprises.
3. People should collaborate using the data to solve plant or company problems together. Remove all spreadsheet silos, so that people do not create and hoard valuable reports. It is also key that everyone has visibility and access to the data, so that people can immediately solve problems rather than debate who has the correct data.
4. The team must discuss EIDI visualization design with those who will be viewing the information, so their needs are heard and taken into account when designing role-based dashboards. The goal should be to make sure they get the information they need quickly and in an easily consumable way.
5. The team should regularly meet with corporate management to keep them engaged and updated on EIDI deployment status. The team plans to produce summarized, easy-to-read key performance indicator (KPI) dashboards that contain fleet status information, such as actual production versus targets, cost variances, and plant availability information. This information needs to be visible to management and others. Management should be kept aware of what gains and cost avoidances have been realized since using the EIDI system.

The Humans behind the Data: Visualization and Collaboration 111

Who Will Be the ProcIndustries EIDI Users?

Bill continued: "Most everyone using the data will need to collaborate. Some will optimize existing refinery operations, others will innovate and improve the processes. This will be a new role for the organization. We need to be proactive about developing innovations in our process operations. We'll categorize each type of user so people know their roles." Bill had the group look at the current ProcIndustries organizational chart (see Figure 1.1 in Chapter 1). Peter Argus, the ProcIndustries continuous-improvement manager, classified the three types of users and their responsibilities.

1. **Subject matter experts and technical users** will be comprised of process operators, maintenance workers, and plant supervisors. The business will benefit from the proposed visualization displays. Real-time alerts and notifications about events that signal a problem or potential problems will provide more time for personnel to make decisions that avoid catastrophes. This helps us better achieve daily targets, resolve immediate issues, maintain production schedules, keep assets healthy, meet daily economic plans, and maintain an environmentally safe operation.
2. **Management users** will be comprised of the production and operations management teams, including local, regional, and division managers. These users will be using the system on a daily basis to assess overall performance of the refinery and ensure that the refinery is running within target ranges. They will define unit performance based on production output, consumption metrics, and product-quality measurements. Management teams will be informed more quickly, rather than receiving after-the-fact reports. These managers will meet with the operational intelligence team to improve operations, review forecasts, and discuss cost projections.
3. **Expert and advanced users** will be comprised of the process improvement team. These users are essential for a continuous-improvement program that seeks to build operational excellence. These users are typically local process engineers, production superintendents, and on-site and remote center of excellence subject matter experts. Their objectives are to detect process or equipment excursions and develop the analytics to configure online performance equations and notifications. They maintain process stability by implementing condition-based equipment and control-monitoring diagnostics. These users should work with the planning

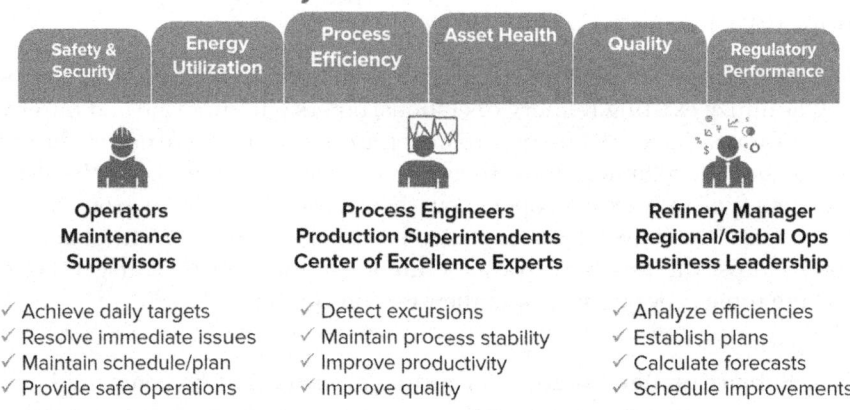

FIGURE 5.1
EIDI user roles and responsibilities.

and economics team to improve scheduling activities of the refinery. The teams can then update refinery production forecasting models to make them more accurate. Figure 5.1 indicates the roles and objectives using EIDI data.

The Impact of Change

As an EIDI system or other cross-team initiative of this scope is deployed, it is critical that the management answers employee questions such as:

- What are we changing?
- Why are we changing our current processes and systems?
- How will the changes impact me?
- What will I be expected to do differently?
- How will I be prepared to handle the change successfully?
- What are the implications of not changing?
- When will the changes occur?
- Will I receive proper training to function effectively?
- Where do I turn if I need help or have questions?
- Is my supervisor or manager on board with this change?
- How can I communicate issues or opportunities for improvement after the change?

The Humans behind the Data: Visualization and Collaboration 113

When planning workflow and responsibility changes, it is important to consider employee concerns. In one of the weekly EIDI transformation team meetings, Bill Roberts provided a summary of the work by Knoster et al. (2000), which provides a framework for thinking about organizational system change. Peter advised the team to review it and give feedback at the next meeting.

Rome Wasn't Built in a Day

Bill Roberts explained to the EIDI planning team that, based on his prior experiences in other companies, this initiative is not a typical corporate project, finite in scope and duration. It is an infrastructure investment that will be a continuous journey toward operational excellence. Bill addressed the team: "We likely won't achieve all of our goals in one or even two years. We think we can roll the first system out pretty quickly, with the right design, planning, and motivated people deploying and using it as we intend." He continued: "Management has put some thought into what the business priorities are and what benefits we can reasonably expect to realize in the near term, while planning to accomplish others as we move forward. We need everyone in the company to understand that we won't accomplish everything right away, but we will set goals and anticipated milestones for this system to assist us in meeting all of our improvements over time. So let's think value now—value over time!"

The five initial EIDI use cases are:

1. Consolidate operations and production data for improved refinery operations, facilitating a solid foundation for situational awareness;
2. Provide quicker access to information by creating useful visualization dashboards, displays, and reports;
3. Implement proactive asset health monitoring in order to reduce unscheduled equipment and process downtimes;
4. Increase product quality, achieve higher yields, lower utility costs; and
5. Improve operational health, safety, environmental monitoring.

The next set of use cases will be:

1. Model the process using a digital twin,
2. Predict failure states and end of life for key pieces of equipment,
3. Holistically debottleneck process constraints, and
4. Perform big data analysis for business improvement.

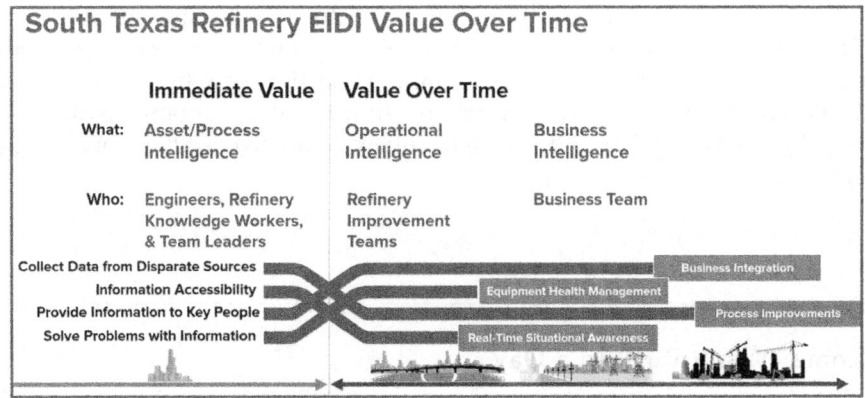

FIGURE 5.2
Proposed near-term and long-term value initiatives.

Figure 5.2 shows the benefits over time using the EIDI intelligently.

The following are key activities that can be performed directly using the EIDI to accomplish business and operations improvements:

1. Real-time situational insights and problem solving through timely visualization
2. Historical data analysis for root cause analysis, process improvement, reporting, and general operational intelligence
3. Real-time analytics and notifications to automatically determine when conditions are unusual or deteriorating

The following are key activities not directly performed using the EIDI but using EIDI contextualized data as input to other software not directly connected to the EIDI (these methodologies are expanded in Chapter 10):

1. Offline engineering, process modeling, operational analytics, and creation of a "digital twin"
2. Extracting highly contextualized data for input as data tables to offline, non-engineering, visualization systems, such as Tableau, Tibco Spotfire, and QlikView, or for input to predictive analytical tools, such as machine learning, deep learning, Python, R, Microsoft Power BI, Amazon Web Services, and other advanced statistical software systems

Peter then described the roadmap the team will follow to embark on Bill's "value now—value over time" mantra, recommending the following path to operational excellence (Figure 5.3).

Phase 1. Refinery operating data is provided to everyone, so they can immediately start solving process problems and reducing production bottlenecks. In this phase, key asset equipment will be

The Humans behind the Data: Visualization and Collaboration

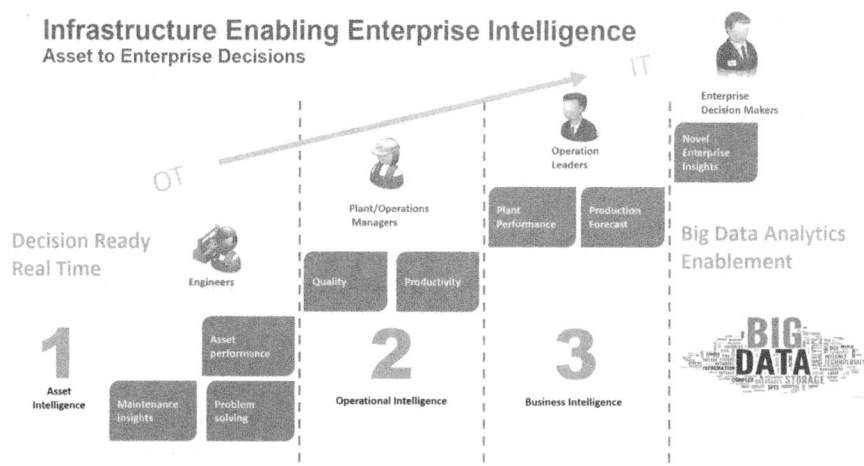

FIGURE 5.3
EIDI roadmap to enterprise operational and business intelligence.

monitored in real time, so that maintenance people are aware where best to spend their time and resources. When operating conditions drift out of optimal ranges, the appropriate people will begin receiving real-time alerts.

Phase 2. Event frame analysis is introduced, so personnel can identify and select key time slices when troubleshooting problems and tracking production. Quality measurements will be integrated into the EIDI on an event basis, so that personnel can identify and determine how process variability affects product quality and yield. Real-time production and maintenance reports are automated to describe variances from expected targets.

Phase 3. KPIs are utilized so that workers become familiar with how they are calculated and how they are consumed. Personnel begins to look holistically at refinery performance, such as electricity and water usage, making sure raw material use is optimized and process targets and distributed control systems (DCSs) and supervisory control and data acquisition (SCADA) are optimally tuned. A center of excellence is formed in this phase to aid in digesting the real-time information and make recommendations to plant personnel to assist them in meeting their production targets and optimizing asset health and safety.

Phase 4. Offline, cloud-based business intelligence (BI) tools and big data analytics are used to optimize business operations and control operating costs. These BI systems consist of statistical tools, machine learning and deep learning, and creating a digital twin representation of refinery operations for simulations and process models (see Chapter 10 for more details).

People-Driven Benefits from the EIDI Implementation

Why is there such an emphasis on the people aspects of the EIDI project? Peter summarized eight people-related benefits that the EIDI can bring to ProcIndustries, once implemented and people are trained to use it:

1. **Motivated staff.** People are more inclined individually and when working in teams to find solutions to issues that make a difference, such as reducing unnecessary costs or improving safety and product quality.
2. **Improved profitability.** ProcIndustries has traditionally utilized a bonus program, based on realizing company profitability and team goals. If the EIDI brings about more cost savings, then it becomes possible that employees can share through larger bonuses.
3. **Better trained, skilled staff.** With easier access to operations data, people can start making decisions based on what the refinery data is telling them, rather than using opinion-based practices.
4. **Less wasted time.** Significant time spent collecting data from various silos will be drastically reduced. People will now spend that time solving challenging issues.
5. **More efficient use of resources,** including equipment, energy, and raw materials. The real-time situational awareness provided by the EIDI enables proactive monitoring of assets and equipment, so that financial resources are more wisely spent. This means servicing equipment when needed, rather than using traditional calendar-based maintenance schedules. Additionally, the use of water and power is optimized by treating them as controllable variable costs, rather than as after-the-fact fixed costs and simply paying the monthly bill.
6. **More timely delivery of work.** By reducing or eliminating unplanned equipment downtime and reducing process inefficiencies, the refinery can better meet production targets and schedules.
7. **Greater ability to delegate tasks.** With better real-time visibility, tasks can be assigned to workers with the right skills.
8. **Improved service** to internal and external customers. Having the data at one's fingertips will reduce the time spent collecting information and responding to internal and external requests.

"Not a bad list of benefits, but not complete," said Bill. "While our people will be better equipped and the business will improve profitability, many of our people remain skeptical." He added, "They argue that they are too busy to do all this training, pointing out they need more time or additional assistance. We have to show them why it is worth it."

Getting the Visualization Right

The team prepared examples to share dashboard design with their peers. Peter explained that one of the new competence center roles is to define best practices and meet with knowledge workers and EIDI system users to hear their views on what they need to accomplish their objectives.

After numerous brainstorming sessions, the team decides to use the following visualization methodology:

1. Create role-based dashboards for getting appropriate information to people where they can easily consume it.
2. Knowledge workers will have ongoing input on what they need to solve issues.
3. Workers receive sufficient training to utilize the system effectively and efficiently.
4. ProcIndustries will allow workers to see any information they need to do their jobs, including process block diagrams, flow diagrams, costs, and so forth.
5. Management will empower people to make decisions based on what the data is telling them.

The team spent the next week thinking about how best to configure and deploy impactful dashboards and displays:

- Summarize the information so the key data is instantly recognizable. Add the capability to drill down into specifics for plant operations people, so they can see how the process is performing and what is causing excursions. Display production metrics so that plant operations people are aware if they are meeting targeted goals.
- Create real-time production KPI dashboards for refinery plant management and ProcIndustries' corporate management, with the ability to view past performance using the time frame of their choice.
- Create real-time and historical unit-based displays for process engineers, with drill-down capability, so they can quickly determine if their unit(s) are running optimally. Data export capability must be provided, so that they can output data to offline calculations, machine learning models, Python, R, and other analytic tools they need to run the units optimally.
- Offer real-time role-based KPI dashboards to plant and corporate personnel with drill-down capability to examine further or troubleshoot issues:
  - Production and operations
  - Maintenance

- Process improvement
- Safety, health, environmental
- Product quality
- Refinery management
- Corporate personnel

Which visualization tools are most suitable to use? EIDI users typically use the following tools, depending on the task at hand and sometimes the person's role and background, as shown in Figure 5.4.

- Process graphic displays, historical trends, and real-time KPI dashboards. These work best for operations and production personnel, process engineers doing root cause analysis, maintenance workers, and anyone wanting to view time trends or graphically formatted material where they can drill down to further access more information.
- Excel spreadsheets are desirable for financial people, doing predefined or ad hoc reports, calculating statistics, or exposing the data to other software programs that use Excel, such as MATLAB.
- Business analytics visualization tools are appropriate for workers performing offline analytics that integrate different types of data other than plant operations data.

Figure 5.5 represents the types of visualization tools with context for using them.

EIDI Users Have Different Needs

Process Engineer
"Can we increase the overall yield?"

Control Room Tech
"The process is like a baby – you have to watch it."

Production Manager
"What is the forecast of productivity?"

Data Scientist
"Can we find new savings with machine learning?"

Reporting Analyst
"I need to combine data from 3 sources in 1 report."

Maintenance Engineer
"I need to know the moment it goes out of tune."

FIGURE 5.4
Data visualization needs for typical roles in a refinery or process manufacturing company.

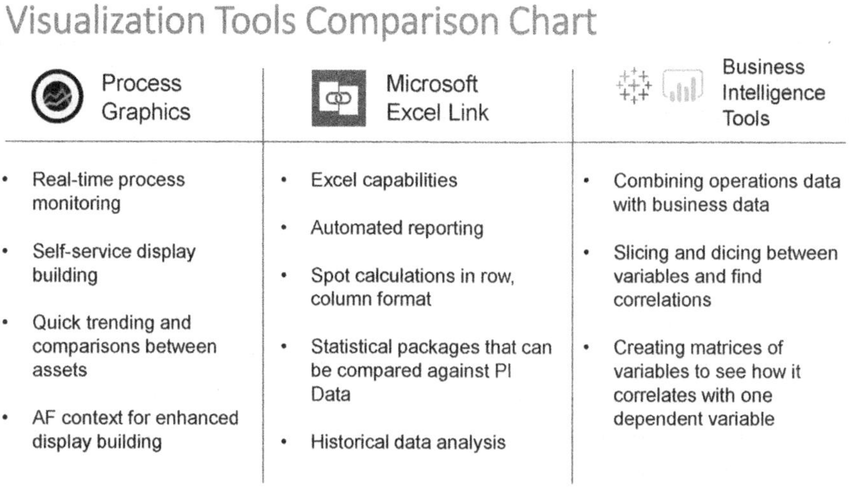

Visualization Tools Comparison Chart

Process Graphics	Microsoft Excel Link	Business Intelligence Tools
• Real-time process monitoring • Self-service display building • Quick trending and comparisons between assets • AF context for enhanced display building	• Excel capabilities • Automated reporting • Spot calculations in row, column format • Statistical packages that can be compared against PI Data • Historical data analysis	• Combining operations data with business data • Slicing and dicing between variables and find correlations • Creating matrices of variables to see how it correlates with one dependent variable

FIGURE 5.5
Types of appropriate visualization tools.

Best Practices on Process Graphics and Ways to Present and Share Information

"With all the data that we have available, the amount of information to present in graphics can be overwhelming," Peter said. "As such, we need to have a hierarchy and a presentation plan on how to go about it."

One of the first objectives is to make sure the data is accurate for the proposed dashboards and process displays. First, the data will be captured, then the data is validated to ensure that sensors are working and sending data properly. Next comes further validation to ensure that the data is within operating limits.

Later in the EIDI implementation, offline tools can be used to holistically monitor the process. If the data is within operating limits (sometimes called *operating envelopes*), physical or empirical models can be used to generate estimates based on current measurements. Process models can be used for this purpose. Because the EIDI system can display future-time projections indicating what values these models are predicting, it will help operators see variation in actual versus expected results. This last step will not be deployed during the initial phases of the EIDI initiative.

It always a good practice to define end users within a system. Who will be using this data, and what is his or her job? The vital ingredient of an infrastructure is that the person in each particular role is the most important element. "In essence," explained Peter, "we are building a cockpit with a data dashboard to simplify the job." Employees should be able to draw a picture or utilize an existing picture from a library. They should be able to easily access trend data so that they have an immediate understanding of refinery conditions. However, people can easily be distracted by color and movement. Flashing colors in abundance become distractions and may cause refinery operators or maintenance workers to lose focus.

Three principles apply to real-time data visualization:

1. **Clarity**
   - Graphics are easy to read and intuitive.
   - Graphics show the process state and conditions clearly.
   - Graphics do not contain unnecessary detail and clutter.
   - Graphics convey relevant information, not just data.
   - Information has prominence based on relative importance.
   - Indications of abnormal situations are clear, prominent, and consistently displayed.
   - Graphics make people aware when information is not being updated.
2. **Consistency.** The colors used by the background and the real-time graphics follow standard, consistent norms, such as green reflects normal/desired range values; red values are an alarm or undesirable condition; flashing values need immediate attention. Consideration needs to be given so that color-blind people can still effectively use the displays.
3. **Feedback.** This refers to the active interaction of a touchscreen or mouse and additional access to detail information. An approach to avoid graphic cluttering is to hide detailed information until the user clicks on an object. However, if the user wishes additional information, the human interface will respond with configured feedback as embedded information in the graphical visualization objects. For example, if you touch a specific trend, additional information such as the current value, minimum, maximum, and rate of change for the selected time period is displayed.

For dashboards, a recommendation is to locate the key metrics on the top, then follow with trends or charts representing the data. If required, present a table with summarized data. For process flow diagrams, it is important to show the data starting at the left of the process supply chain and ending at the right with the results.

Mobile Access to Information

South Texas refinery production manager Tim Olsen said he had been exploring the EIDI's PI Vision displays using his iPhone when away from the refinery. Using secure access mechanisms, this mobile capability enables people to access displays using their phones, tablets, home PCs, and laptops

Mobile Device Support for Collaboration Anywhere

Requirements:

- Easy to use, self-service, and scalable
- Access data from any web browser, including mobile device browsers
- Organize and share displays across the ProcIndustries organization
- Enter EIDI data manually from mobile devices

FIGURE 5.6
Accessing EIDI data from mobile devices.

(Figure 5.6). He said he was able to view a trend display showing conditions of the boiler house. "I was never able do that before," declared Tim.

Asset-Relative Displays

When a company such as ProcIndustries embarks on an EIDI project, there seems to be an infinite number of ways to display the process, equipment, and production metrics. One advantage of the method chosen by the ProcIndustries team is to build an asset data model, used to establish a hierarchal structure and to simplify system configuration. This is accomplished by using the EIDI system's asset framework (AF) modular templates that enable configuration of similar assets (pumps, compressors, distillation columns, etc.) using a single, reusable template. When building the asset template, the following items are included:

- The individual variables (data streams) that comprise a specific asset;
- Online, run-time calculations and alerts specific for this specific type of asset (perhaps an inferred efficiency measurement is calculated using measured sensor values. A real-time alert or notification is then generated when the efficiency measurement result falls outside acceptable limits.); and
- Other reference data that may provide helpful information but not used in real-time processing, such as when the equipment was placed into service or when it was last serviced, its typical maintenance cycle, or the vendor-recommended operating range.

Using the AF modular template approach, the team discovered they could build a single visual display for similar units and simply navigate to select the desired unit to view its status and supporting data.

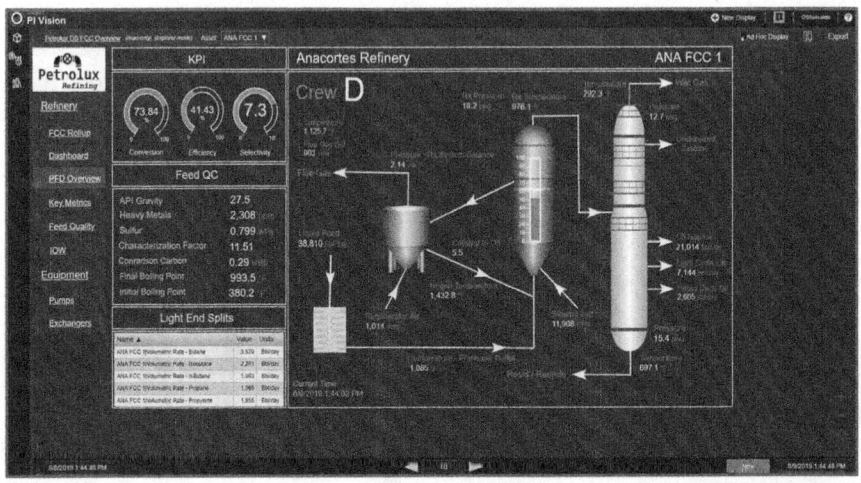

FIGURE 5.7
Navigating through refinery units using asset-relative displays.

Figure 5.7 shows this capability when displaying a refinery unit, with an AF tree view on the left side of the display for navigating through the refinery's units. These displays can show process and quality variables. They can use statistical process control (SPC) charts to plot quality data with the SPC analysis tools. This display also shows a y versus x linear plot of two process variables such as the effect of the gas flow rate on the reactor discharge temperature. The user decides what to include for each unit, based on their needs. They can modify the base display to add or modify variables through configuration tools. This is an invaluable aspect of the EIDI. This strategy reduces the cost of curiosity by empowering people to resolve problems.

Process Improvement through Visualization

For process engineers, the EIDI provides real-time and historical data that is vital to troubleshooting and improving the refinery's unit processes. Plant engineers can perform many simple debottlenecking efforts through efficient use of trend displays, specialized charts, and event analysis. The EIDI real-time data management system provides standard, configurable visualization tools for many of these needs. Following are some examples:

- Performance trends show targets and current measurements used for root cause analysis (Figure 5.8).
- Holistic dashboards and refinery-wide displays use drill-down capability to unit-specific data.
- Event-framed displays compare event activity against other events or against an ideal event.

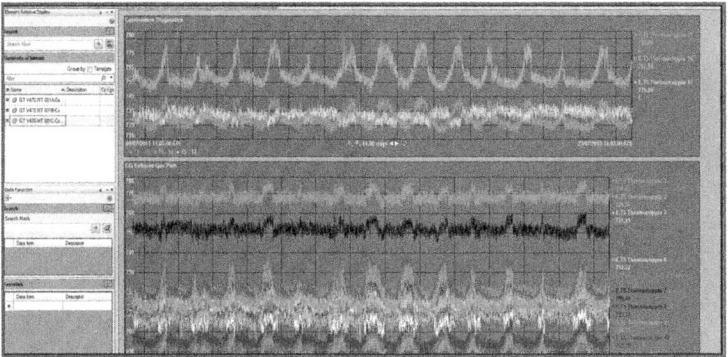

FIGURE 5.8
Real-time data trend display for root cause analysis.

- Metric predictions are used in variance analysis, comparing current versus expected performance.
- Statistical quality control (SQC) and SPC charts can be used. SPC helps ensure that the process is operating normally (Gaussian distribution). If the process moves more than 3 sigma in variance, automatic alerts will be generated because the process is drifting outside its normal variance. There are a number of statistical methods to model the processes and to define faults if the process has deviated from normal, expected ranges.
- Linear (x–y) and scatter plots show the results of two time-based data sets, so that a pattern can be deduced.
- Actual performance is tracked in comparison to expected or predicted results (e.g., process model predictive data).
- Data is classified and aggregated using Pareto charts, displaying cumulative totals for priority analysis.

The refinery engineers and other knowledge workers were then able to extract data, such as the example trends shown in Figure 5.8, and import that and any other EIDI data into Microsoft Excel (Figure 5.9). The EIDI menu in Excel provided a variety of options to populate their spreadsheets with current values, historical archived values, derived values, or statistical data on the archived values. Figure 5.9 demonstrates an example of preformatted or ad hoc Excel reports.

Some visualization tools, such as multivariate charts and fishbone diagrams are deployed outside the real-time data management system, but can leverage EIDI data, such as

- Cause-and-effect trends (fishbone diagrams);
- Multivariable charts (key metrics vs. manipulated variables); and
- Pie charts, bubble charts, and other business-oriented charts.

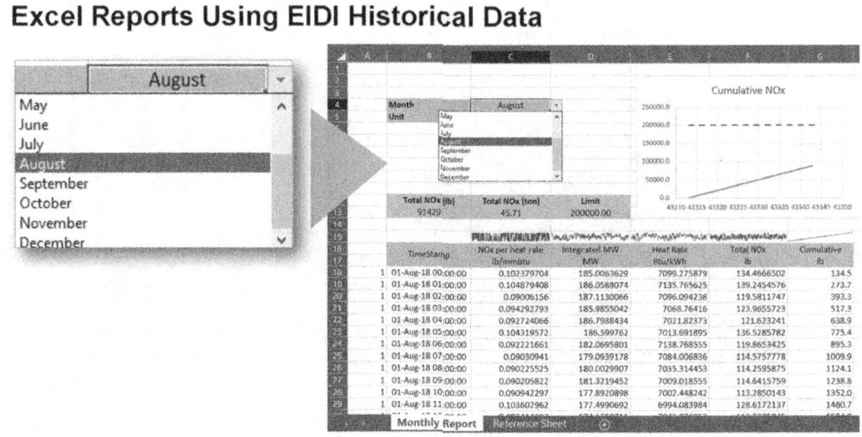

Excel Reports Using EIDI Historical Data

FIGURE 5.9
Creating Excel reports using EIDI data.

Using analytics to create these visualizations, refinery operating people can embed acceptable operating ranges and monitor real-time behavior, using real-time notification alerts when the process strays outside preconfigured upper and lower control limits. This is generally an early warning indication that the process unit is in trouble. Concurrently, there are mechanical limits used for equipment monitoring. The goal is to integrate production and maintenance strategies for optimal performance, that is, goals of no excess process deviations and no unscheduled downtime events. These displays and dashboards are available on desktop computers, laptops, or on mobile devices (tablets, smartphones) connected to the industrial network via tools like a virtual private network (VPN) or other secure access mechanism.

Creating Dynamic Performance Operational Displays

Peter told the team that users are able to create their own specific data visualizations to help them perform their jobs easily and more quickly. Any authorized user can display EIDI data at the desired level of detail. The visualization tools examine process and asset data based on current time and/or in their desired time context, which is vital for forensic cause-and-effect analysis.

Monica Armstrong, the planning and economics coordinator, explained that cloud computing makes access to these applications more flexible. It is now possible to access and create real-time graphics or use traditional Microsoft Excel or Power BI analytics tools on-premises or via the cloud. End users can create individualized KPI dashboards and access them from their workstations or from a smartphone or tablet.

"This ability to include only the data that you need has another benefit," said Monica. "It means you don't waste any time viewing unnecessary refinery information. You can focus on what is relevant, and if needed, drill down into the data you are interested in."

Peter added that the visualization tools let people monitor operations at different levels: the process unit, for example, or the refinery complex, or at the enterprise level. Users can select standard displays that the EIDI digital transformation team has configured or they can modify displays creating the context they desire.

Three activities that enable workers to create and use EIDI dashboards and graphics are authoring, monitoring, and ad hoc analysis. The EIDI team will initially help users by publishing a set of predefined display and dashboard templates for the users. If knowledge workers wish to customize their dashboards and displays to make it easier to consume the information, the EIDI team will train them how to customize and publish their own displays. The second activity is to train users how to visualize and navigate through displays and dashboards. The third activity is for the team to train power users so that they can effectively design customized displays and dashboards to solve problems unique to them. Figure 5.10 shows some examples of this via typical EIDI dashboards or other applications that use EIDI data.

"The EIDI is set up to serve as the primary data access vehicle," said Peter. The goal is to have engineers, a planning and economics manager, or the maintenance team configure their dashboards, to solve problems themselves, not needing to wait for anyone to generate queries or reports.

"If they need assistance to interpret the data, it's OK," Peter said. "The key is to embed the EIDI in our daily work routines. Then we will be alerted when a part of the refinery, whether it is a piece of equipment, or a chemical mixture or some other factor, is becoming a problem, either an asset misbehaving or production not meeting targets." (See Figure 5.11.)

FIGURE 5.10
Collaboration activities for knowledge workers.

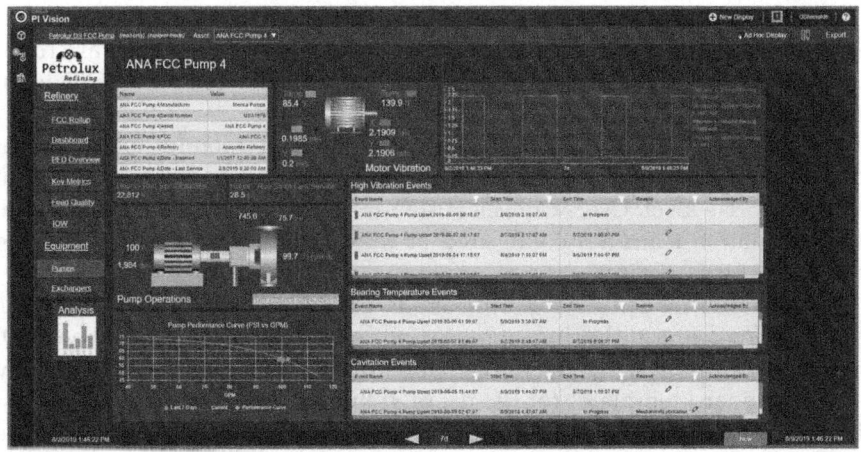

FIGURE 5.11
Display showing compressor status and performance.

Data Turns into Workflows: Workflows Adopted by People Turn into Meaningful Change

As the South Texas refinery continues to implement the EIDI, and the work teams start to use it as part of their daily work routine, the transformation team notices that people are starting to use the EIDI data to address issues within their own work groups. Once they find they have achieved some initial successes, they start to collaboratively work with other teams at the refinery, armed with the needed information.

With success, they begin to see challenges unfold that must be addressed. Change management is one of these issues. Improvement means change, and change means a different way of working. As processes and systems change, people often need to change their way of doing things. This is difficult, as most people have a natural resistance to change. Change is also difficult for the people who design systems. When implementing change, they sometimes try to replace people, limit their actions, or avoid human involvement. This tendency limits the effectiveness of the resulting systems and processes. Often companies forget that there is a natural reduction in productivity during the implementation of systems or workflow changes. Not allowing for this tends to lead to longer periods of system underutilization.

Management often deems an initiative of this magnitude prematurely complete when initially realizing quantifiable positive results, when a number of the old problems are resolved. In reality, this is the time to assess workflow effectiveness and make changes where necessary. It is also a good time to brainstorm and determine how the system and workflows can be better improved to solve additional business problems or reduce worker inefficiencies.

What You Should Take Away

In this chapter, we learned that management needs to fully endorse the EIDI initiative, support employees in their adoption of the new system, and make sure the team has sufficient training to be successful utilizing the new system. Leadership should encourage collaborative efforts in solving business issues and should reward people demonstrating that behavior.

If there are change management issues, either logistical or with people struggling with the change, the team leadership needs to address and solve these issues.

ProcIndustries has taken the appropriate steps to make sure refinery teams' voices are heard so that the new EIDI and new workflows address everyone's needs. They have also put a plan in place to make sure people are properly trained how to access and visualize EIDI data. Employees will receive sufficient training so that they can customize their personal dashboards and displays to fit their needs.

The teams are getting more comfortable in collaborating among the various refinery groups, as evidenced by more people following the new workflow guidelines when they detect an abnormal operating condition. Before long, the engineers will start to investigate and recommend how to improve plant processes, based on historical data analysis. The maintenance team has already begun to implement a proactive condition-based maintenance program to identify the underperforming refinery assets. In the next chapter, we present strategies for asset management best practices.

Reference

Knoster, T., Villa, R., and Thousand, J. 2000. "A framework for thinking about systems change." In *Restructuring for Caring and Effective Education: Piecing the Puzzle Together*, ed. R. Villa and J. Thousand. Baltimore, MD: Paul H. Brookes. pp. 93–128.

Additional Reading

Bascur, O.A. 1990. "Expert process operation advisor." In *Control 90': Mineral and Metallurgical Processing*, ed. R.K. Rajamani and J.A. Herbst. Littleton, CO: Society for Mining, Metallurgy & Exploration. pp. 67–76.

Bascur, O.A., and Halhead, M. 2013. "Energy effectiveness and sustainability management at Anglo American Platinum." In *Proceedings Copper 2013 International Copper Conference*, vol. 1, ed. C. Moscoso, J. Rosales, and A. Vio. Santiago: Chilean Institute of Mining Engineers (IIMCH). pp. 415–423.

Bascur, O.A., and Kennedy, J.P. 1995. "Measuring, managing and maximizing performance in petroleum refineries." *Journal of the Japanese Petroleum Institute* 102.

Bascur, O.A., and Kennedy, J.P. 1996. "Measuring, managing, maximizing refinery performance." *Hydrocarbon Processing* 75(1):111–116.

Bascur, O.A., and Soudek, A. 2014. "Strategies for implementation of energy effectiveness and sustainability: Example Anglo American Platinum." Presented at the 12th AusIMM Mill Operators' Conference: Achieving More with Less, Australasian Institute of Mining and Metallurgy (AusIMM), Victoria, Australia.

Brynjolfsson, E., and McAFee, A. 2014. *The Second Machine Age: Work, Progress and Prosperity in a Time of Brilliant Technologies*. New York: W.W. Norton.

De Geus, A. 1988. "Planning as learning." *Harvard Business Review* (March/April).

Duggan, K.J. 2012. *Design for Operational Excellence: A Breakthrough Strategy for Business Growth*. New York: McGraw-Hill.

Fogler, F.S., and LeBlanc, S. 1995. *Strategies for Creative Problem Solving*. Upper Saddle River, NJ: Prentice Hall.

Hollifield, B., Oliver, D., Nimmo, I., and Habibi, E. 2008. *The High Performance HMI Handbook: A Comprehensive Guide to Designing, Implementing and Maintaining Effective HMIs for Industrial Plant Operations*. New York: Vanguard Press.

Kresta, J., MacGregor, J., and Marlin, T. 1991. "Multivariate statistical monitoring of process operating performance." *Canadian Journal of Chemical Engineering* C9:35–47.

Lieberman, N. 2009. *Troubleshooting Process Operations*, 4th ed. Tulsa, OK: PennWell.

Markman, A. 2012. *Smart Thinking: Three Essential Keys to Solve Problems, Innovate, and Get Things Done*. New York: Penguin.

Miller, A. 2014. *Redefining Operational Excellence*. New York: American Management Association.

Pietersen, W. 2002. *Reinventing Strategy: Using Strategic Learning to Create and Sustain Breakthrough Performance*. New York: Wiley.

Rummler, G.A., and Brache, A.P. 2013. *Improving Performance: How to Manage the White Space on the Organization Chart*, 3rd ed. San Francisco, CA: Jossey-Bass.

Turton, R., Bailie, R.C., Whiting, W.B., and Shaeiwitz, J.A. 1998. *Analysis, Synthesis, and Design of Chemical Processes*. Upper Saddle River, NJ: Prentice Hall.

Woods, D. 1994. *Problem Based Learning: How to Gain Most from PBL*. Hamilton, ON: W.L. Griffen Printing.

Woods, D. 2005. *Successful Trouble Shooting for Process Engineers: A Complete Course in Case Studies*. Weinheim, Germany: Wiley-VCH.

Zhang, Y., Vaculik, V., Dudzic, M., Miletic, I., Smyth, A., and Holek, T. 2003. "Industrial application of multiway PCA to continuous casters." In *IFAC MMM Workshop Proceedings*. Laxenburg, Austria: International Federation of Automatic Control (IFAC).

6

Preventing Abnormal Situations

One event can be the cause of another only if they both can be brought within the same point of space.

Andre Breton

The scarce resource never was technology; it was the set of managerial capabilities needed to create value with that technology.

Vijay Gurbaxani
Professor of Information Systems, UC Irving

Chapter Overview

This chapter describes two important areas where the South Texas refinery improves its operations to achieve best practices: asset management and their evolution from regulatory control to smarter integrated operations. You will see their transition from reactive, corrective maintenance practices to a proactive, condition-based preventive maintenance methodology. In the past, the refinery often performed calendar-based servicing of critical assets, which often did not align with degrading asset performance behavior. As a result, the refinery experienced unscheduled shutdowns caused by faulty equipment. The second consequence was that the refinery often serviced equipment too frequently, resulting in unnecessary costs, simply because the equipment supplier recommended servicing at specific intervals. This chapter describes how the refinery implements a condition-based maintenance program using the enterprise industrial data infrastructure (EIDI), where the actual performance data determines which maintenance activities need to take place. As a result, maintenance service costs and excess labor costs are reduced. In addition, industry best practices are presented for predictive maintenance, which estimates time-to-failure for critical assets, based on historical performance data.

You will also learn about the evolution of process control effectiveness. The overall strategy of the digital transformation team is to have employees spend less time handling mundane tasks by encouraging and supporting task automation. The plan is for employees to resolve problems they can, rather than having to search for relevant data and debate with others regarding which data is accurate. Empowering workers to make quicker decisions that impact plant operations and production is critical to effective business operations.

Peter Argus, the continuous-improvement manager, is looking at these new strategies with Pat Verlaine, the plant's information technology (IT) manager. Pat is supportive of having the data available to employees using new dashboards and collaboration-enabling tools, such as online EIDI dashboards, displays, and trends (OSIsoft's PI Vision), Microsoft's Power BI and Machine Learning Studio, Anaconda Spyder for Python, and other analytical tools. There have been several months of productive partnership between Peter and Pat discussing new process improvements.

Peter and Monica Armstrong, the plant's planning and economic coordinator, are satisfied because they know that the plant's planning and operational strategies can be aligned with the use of modern technology.

ProcIndustries currently has been struggling to satisfy business opportunities with their rigid legacy system. Departmental coordination has been a challenge. Each functional area has been using different systems with inconsistent views of their assets. As a result, most employees spend excessive time seeking data for problem solving. They typically exchange emails or phone calls, checking for necessary information. Their current systems were implemented to support independent functional environments and facilitated working in silos as opposed to working collaboratively. Their computers and networks now allow them to integrate information into a single system of record at the refinery.

Operations and maintenance teams have somewhat different objectives and are often limited by legacy systems. They lack dashboards or access to comprehensive real-time production and equipment information.

Real-Time Data Analytics to Improve Operational Support

Peter Argus meets with Chuck Smith, the process control engineer, and Tim Olsen, the production manager, to develop strategies that enhance equipment performance, process control monitoring, and energy management and create production improvement strategies. Peter explains that integrating these new digital operational strategies enhance support systems. Chuck shares a diagram (Figure 6.1) presenting real-time operations and enabled support systems using the EIDI, which he calls the "business objectives pyramid."

"This is a good place to start the discussion. A system of record, storing real-time data and events, can serve many functions in our business. As we have seen, transforming data into operational insights for reducing plant problems is necessary. These hidden losses are there. We need to reduce abnormal situations to increase overall production effectiveness," acknowledged Tim.

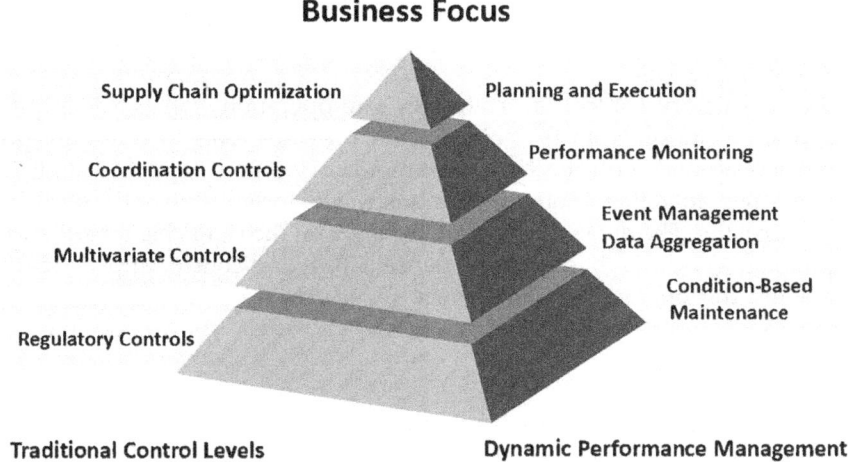

FIGURE 6.1
Increasing the scope of industrial plant for process control and operational intelligence.

Business Objectives Pyramid

Proper application of today's technologies enable the process industries to develop competence centers to manage their industrial complexes, integrating operational and business objectives.

The Left Side of the Pyramid

The left side of the pyramid in Figure 6.1 shows traditional engineering control levels fed by sensors and data collected from the processes. You may think of it as operations technology or OT. At the lowest level is the instrumentation level, which consists of devices for acquiring data from sensors, field displays, and hardware safety interlocks for ensuring safe emergency shutdowns. The instrumentation level sends the data to the regulatory control level, which consists of traditional control hardware such as distributed control systems (DCSs) and programmable logic controllers (PLCs). This level integrates real-time data for the regulatory control level. It is one of the most important levels because it has to be extremely robust and responsive for industrial process continuity and operational safety.

Layer 1: Regulatory Controls

The regulatory controls maintain process variables at their prescribed setpoint and stabilize variations caused by local disturbances occurring at a timescale of seconds to minutes. Various causes originate the disturbances, such as weather conditions, changes in raw material characteristics, and start-up and shutdown at other refinery sections of the supply chain. In addition, this layer allows the operator to take manual control of the plant in cases of emergency.

The regulatory control layer ensures safe operation, collects data at the original resolution from the sensors, and provides tools to configure process displays and control actions. The regulatory control layer transmits the streaming plant data to a dedicated industrial historian, akin to the black box of an airplane. The data stream feeds the layer on the right side for enhanced equipment condition-based maintenance (CBM) and control improvements. As such, the operational intelligence reuses the real-time data with the proper tools for time-series analysis.

Layer 2: Multivariate Controls

Layer 2 evolved because of advances in control algorithms, hardware, and computing capabilities. In a typical processing plant, the problems are generally classified as multivariable, with many control interactions caused by nonlinearities of the process, process equipment constraints, and unknown process disturbances. Because of the possible interactions among the variables, all control moves must be carefully coordinated. The control actions are taken to accommodate longer duration disturbances, usually minutes. This is also caused by slow processing times of online process sensors or instream process analyzers.

ProcIndustries' refinery units are complex, presenting unconventional process dynamics. The process dynamic requires streaming data to develop process models for control design and maintenance.

Layer 3: Coordination Controls

The process coordination controls in layer 3 of the pyramid integrate all the process controls, quality process controls, constraint controls, and the environmental and safety controls for each process unit. These process controls require layer 3 right–side support to analyze the operational data and identify process and equipment constraints.

Layer 4: Supply Chain Optimization

The plant supply chain optimization is implemented to balance overall constraints to find optimal steady-state operating conditions of the plant, based on current production requirements and factors such as raw materials, energy and consumable costs, and production demand. Plant coordination activities are related to planning and scheduling activities reported on the right-hand side of the pyramid. "Imagine if we could improve our daily planning and our overall process coordination," said Chuck. "Monica and Peter showed us how real-time data can be used to detect opportunities for reducing minor losses in the refinery." (See Chapter 3.)

To remain competitive, the team must be vigilant about proactively maintaining equipment to achieve optimum availability and performance levels. *Reusing data* from the industrial data infrastructure is the most cost-effective way of achieving high productivity and performance levels. Ideally, operations and maintenance teams must collaborate, using the same data to keep the equipment online 24/7 for optimal production levels. Using the capabilities of a modern industrial data infrastructure, maintenance personnel can implement real-time condition-based and predictive maintenance strategies. Real-time analytics trigger alerts to troubleshoot problems prior to catastrophic failures.

The Right Side of the Pyramid

The right side of the pyramid in Figure 6.1 represents business applications, big data, and operational intelligence support that can be used to increase the scope of the critical data infrastructure software. Much of this represents the IT side of the business. "Planning and scheduling activities are usually set from integrating the plant industrial data infrastructure with our enterprise business systems, which have production plans and utility costs," stated Chuck.

Layer 1: Condition-Based Monitoring

Layer 1 on the right side of the pyramid depicts the support center, which provides actionable information by all functions to improve the current state of the enterprise. For stable process plants, process equipment must maintain its highest availability in harsh industrial processing environments. Until reliable technologies became available, fully functional condition-based monitoring was not deployed.

Today, reliable process and equipment data streams enable maintenance to assign the necessary context to families of equipment. Thus, the required rules can be set and process analytics can be performed to classify the data to generate alerts and notifications, indicating a possible unscheduled shutdown looming ahead.

Underperforming regulatory controls and multivariable control loops cause process variability that adversely affects profitability. Having the data available in an easily accessible format for advanced analysis simplifies the continuous improvement required to support process controls at all levels for all plant process units.

Layer 2: Event Management

Layer 2 on the right side of the pyramid provides the support to validate and classify the data to develop information needed by continuous-improvement teams. These teams identify opportunities to ensure that all plant equipment is running or available to be run, with communication smoothly flowing among workers. Using online process analytics, the team can derive operational-mode time intervals in a plant and evaluate production and operational costs on a shift-by-shift basis. The transformation of data into actionable information using operational events to obtain production, energy, water, reagents, and other variables at the adequate degree of detail is essential. Transactional systems can be integrated with operations and production data to proactively improve overall production performance of the process plants (Bascur and Aroqui 2014; Bascur and Halhead 2013).

Layer 3: Performance Monitoring

After condition-based monitoring and assessments, performance monitoring (layer 3 on the right side of the pyramid) provides the final results. These results are transmitted to enterprise resource planning (ERP) systems to report all yields, product quality, operating costs, equipment availability, and inventories. Performance monitoring uses the advanced analytics of event data management to disseminate the data into actionable information. At this level, soft sensors can be implemented using predictive analytics, which are explained later in the chapter (Steyn et al. 2018). The physical sensors are backed up by the soft sensors, increasing process control robustness and stability.

Proper data classification and aggregation at the desired level of detail (to calculate all key operational metrics) enables faster communication and collaboration within the functional teams in the refinery. For example, it is always a delicate balance between the operations unit excessively pushing the equipment, and the maintenance team excessively maintaining the equipment. As described in Chapter 4, the data classification capabilities permit online mass balances to predict yield and run sensitivity analysis, thus increasing the value of the entire system (Bascur and Soudek 2019).

Layer 4: Planning and Execution

The analysis of information for overall production profitability, based on supply and demand, are provided by the planning and execution teams (layer 4 on the right side of the pyramid). Standardizing the information enables fine-tuning it for optimization tools. Planning and execution provides targets and schedules for the optimal refinery operations. Having the data infrastructure capabilities in layer 2 enables company personnel to quickly reduce operating costs. It also has enabled cloud-based technology where a production or manufacturing company can directly provide third parties (suppliers, service providers, domain experts, etc.) with highly contextualized data through a secure real-time cloud connection. An additional layer of analysis and support through equipment maintenance, catalyst provider support, water management, and outsourced external support has become a reality and will be further discussed in Chapter 9 (Bascur et al. 2016; Cope and Chugh 2019).

Accessing or sharing data with external services augments the acquisition of knowledge and extended support of remote plant operations. The collection and contextualization of real-time data is essential in providing input to more sophisticated offline predictive analytics and data analysis tools. Using the asset and time context in the EIDI allows for production event generation and provides aggregated data at the desired level of detail.

Enhancing Equipment Availability

Proper maintenance of plant systems and equipment supports optimal plant operation, capacity, and productivity; improves output quality and worker safety; reduces the likelihood and severity of plant and machinery downtime; and pares overall operating costs. Although maintenance covers many facets, the main objective now is to augment the effectiveness of current maintenance strategies by real-time monitoring of equipment health and processes using digital technologies (Figure 6.2).

Asset optimization is defined as the approach to maintenance that allows operating at minimum cost. Maintenance strategies are specific for each piece of equipment and contribute to its process function. The total cost associated with an equipment failure is the sum of the costs related to the availability of the equipment while the plant is down for repair; the costs associated with the repair and maintenance; revenue lost by not supplying product to customers; and increased environmental, safety, and asset risks. A computerized maintenance management system, also known as *enterprise asset management*, is used to coordinate equipment parts inventories, planning, and maintenance policies. The equipment and associated elements are related to plant areas and production lines. Each of these elements are subject to multiple failure modes. For example, ProcIndustries has many processing areas, thousands of equipment pieces, and hundreds of thousands of sensors with different types of failure modes. This is a huge challenge for the digital transformation team.

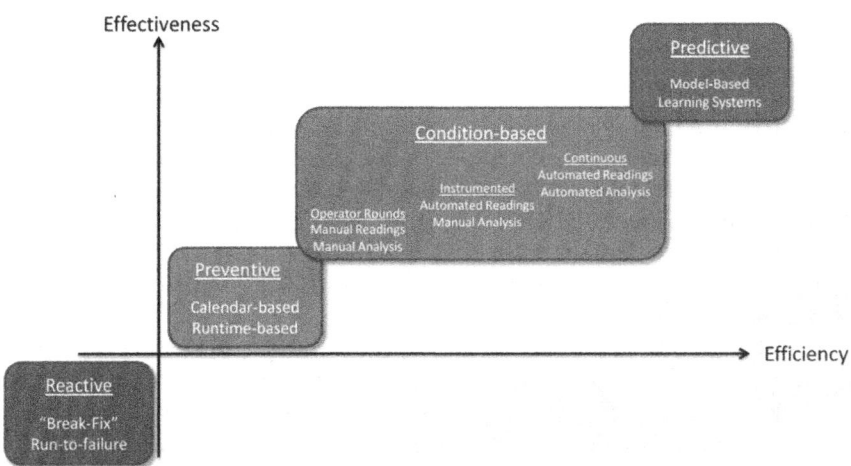

FIGURE 6.2
Typical industrial maintenance strategies.

With raw sensor data and real-time calculations, such as totalized run times, asset performance efficiency measurements, and unscheduled asset downtimes, process, energy, and manufacturing companies can transform their maintenance programs from

- Time-based (servicing equipment every X months) *to*
- Condition-based (continuous monitoring and taking action before a failure) *to*
- Predictive (calculating when an asset is expected to fail) *to*
- Prescriptive operational maintenance (operational modes, production, and equipment) *to*
- Community-based maintenance (integrating remote service providers).

Transforming to a data-driven maintenance strategy allows manufacturers to reduce catastrophic failures, reduce business losses from being unable to supply product, more accurately budget for maintenance and replacement parts, improve safety, and avoid environmental issues.

Reactive Maintenance

Reactive maintenance is a "run to failure" strategy, appropriate for some non-critical assets when impacts and costs of failure are minimum. It requires a minimum of setup and oversight in terms of inputs such as workers and specialized sensors. Usually, there is no preventive maintenance performed on the assets. The decision is to let the asset run until it experiences a failure and then perform corrective (reactive) maintenance. It is often "the squeaky wheel gets the grease approach."

Preventive Maintenance

Preventive maintenance includes calendar-based preventive maintenance and corrective maintenance in response to incipient (or catastrophic) failures. The strategy requires periodic asset maintenance, such as cleaning and inspecting, lubrication, oil replacement, packing bearings, filter checks, and diagnostic checks (e.g., thermography, eddy current testing). This approach often requires that the asset or system be taken off line (or out of service).

This is a conservative maintenance approach that can sometimes lead to induced failures because the dynamic behavior of the asset is not taken into account. It assumes all similar assets degrade at the same rate. Sometimes, it is the only means to maintain the asset due to lack of instrumentation and performance data.

Condition-Based Maintenance

CBM is generally applied to a set of critical assets that have significant repair and replacement costs or cause significant impacts to the business when they fail. Specialized condition monitoring equipment many be required to track the health of assets and respond to trends or events that indicate a degraded condition. However, in most cases, a plant can derive indicators or use soft sensors that provide equipment totalized run times, number of starts/stops, time to transition or operate, excessive temperature, delta pressures across a screen or filter, or drastic variations of power measurements of rotating equipment. In addition, simply calculating the asset or system's efficiency or performance is very valuable.

The ProcIndustries digital transformation team quickly realizes that they can start with the data collected in their EIDI without waiting to deploy specialized sensors. Continuous improvement is their mantra, so they will keep building more sophisticated tools. The significant economic benefits of this transition are shown in Figure 6.3.

Predictive Maintenance

Predictive maintenance is a model-driven monitoring approach that uses both maintenance and operational observations to improve reliability and overall plant operations. This strategy requires that an operational model exists for the asset, system, or process, for a given set of ambient and input values. The model provides anticipated values for process parameters. It is

Condition-Based Maintenance Anticipated Benefits

- Financial - Change the maintenance management cost profile
  - Moving to condition-based maintenance reduces calendar-based - preventive maintenance.
  - Emerging failures are detected before functional failures.
  - The result is more targeted capital expenditures with eventual overall reductions.

- Organizational benefits
  - Codification of organizational intelligence into condition-based algorithms
  - More effective work prioritization
  - Improved decision-making capabilities

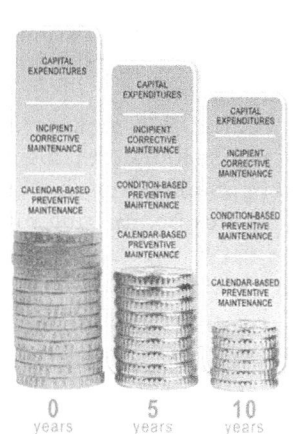

FIGURE 6.3
Expected decreases in outlays when using proactive maintenance strategies.

extremely effective in answering the question "When is the asset or process not operating as it has in the past?"—also referred to as *anomaly detection*. Some industries call it *advanced pattern recognition* (APR).

Detection and diagnosis of abnormal process conditions or critical conditions is essential for the effective operation of ProcIndustries. Fault detection from measured data is typically done by identifying a region of normal process operations and evaluating in real time if it falls outside this operating region. Many algorithms define abnormal conditions using statistical process control (SPC) techniques, multivariate analysis, and machine learning (ML), to name a few. The key is to predict abnormal process conditions using advanced analysis for a certain cause-and-effect situation.

With the availability of faster, more intelligent ML algorithms, predictive maintenance is now becoming feasible on a division- or enterprise-wide scale. Unprecedented opportunities afforded by the industrial internet of things (IIoT) have further changed the playing field, and there are potential benefits yet to be realized. For example, predictive maintenance, originally based on selected asset condition data, has grown to accommodate online, real-time streams of multiple types of condition data received via sensors and even drones. Companies are applying ML algorithms to further refine their predictive analytics and prognostics. Edge computing is enabling process sensor data very close to process units for fault detection and vibration monitoring (Dewald and Santhebennur 2019).

Paul Morgan, the plant maintenance manager, realizes the value of his team's easy access to process and equipment real-time data. He explains, "Predictive asset management is quite a different approach from what we have been doing reactively in the past. A proactive enterprise asset management strategy demands partnership across the enterprise to ensure that all operations act in concert to keep assets working at their optimal level of performance."

Paul is also looking at the ISO 55000 standard by which corporations establish long-term business goals to manage assets. Bill Roberts, vice president of operations, has asked the team to find a way to reduce implementation time by utilizing modular asset templates for similar assets across the refinery: "We can implement in one place and replicate many times for all our sites."

The newest opportunity, prescriptive maintenance (R_xM), is a multivariate approach that merges asset condition data with any combination of operating, environmental, process safety, engineering, supplier, or other related data to better diagnose conditions and prescribe specific options for corrective action. Advanced analytics, pattern recognition, modeling, ML, and artificial intelligence (AI) enables R_xM so that companies can minimize the need for reactive maintenance on critical equipment. In addition, as older, experienced maintenance workers leave or retire, their knowledge will remain through development of these algorithms.

Condition-Based Maintenance: The P–F Curve

Paul shows the team the so-called maintenance potential failure to actual failure (P–F) curve (Figure 6.4): "Many of you may have seen this. I believe this is very important and like to use it to describe a common understanding of equipment failure. Every system or piece of equipment has something that can be detected when its performance is degrading. It could be a slight change in temperature, a high vibration, or a change in amperage. The *P point* on the curve represents the time that change is detected. Every system or equipment will also have a point when it fails; that is the *F point* on the curve. The time between these two points is our opportunity time to avoid a catastrophic failure. The farther up you can move the point of detection, the better opportunity you have to commercially plan around the market and schedule outages at the best financial time, while planning for acquiring parts and scheduling labor. You need proper instrumentation and real-time data availability on your critical equipment to have earlier detection of pending issues and move that up the curve."

Figure 6.4 profiles the equipment condition from initial detection to catastrophic failure. Measures detected that assist in assessing the equipment condition to define an action include ultrasonic detection, vibration detection, oil analysis, audible or sound, temperature change, mechanically

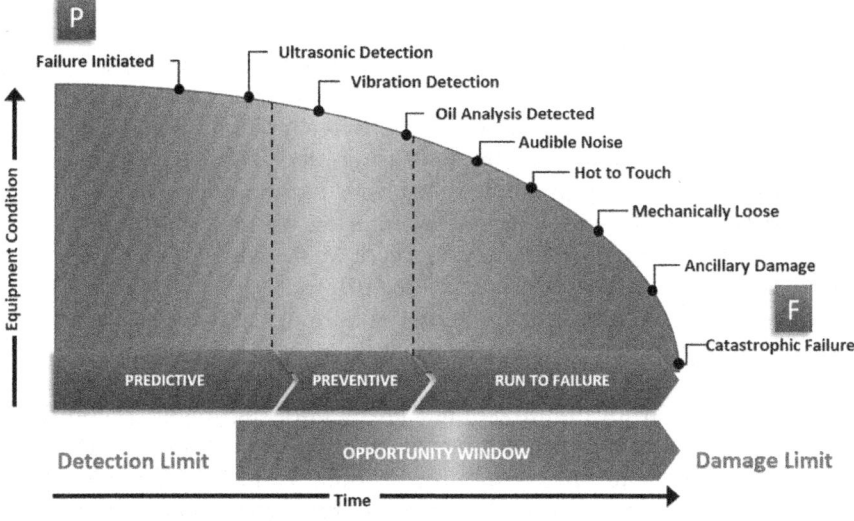

FIGURE 6.4
Condition-based predictive failure P–F curve.

loose, ancillary damage, and ultimately catastrophic failure, resulting in an unscheduled downtime, which can be very costly.

The "run to failure" portions of the graph in Figure 6.3 represent the risk of asset failure assuming the test method indicates a failure is near. The response time varies based on asset, type of test, and frequency of the testing method. Quite often, there is little time to respond, depending on plant conditions, time of the notification, and so forth. Most commonly, alarms for critical items are sent to the operations department to ensure that equipment can be moved into a safe condition prior to catastrophic failure. Although these predictive techniques technically indicate a condition of the asset, they are not often used by a computerized maintenance management system (CMMS) because they are not automatically forwarded to the CMMS.

A simple example of the workflow was presented in Chapter 4: The EIDI generates a notification, which is sent to operators and the maintenance team, who analyze the situation and schedule changing of the part at the appropriate time.

CBM replaces vendor recommended maintenance schedules with activity driven by equipment condition data. It allows planning and scheduling of maintenance activities or repairs before functional failure. With CBM, organizations perform maintenance only when needed to prevent operational deficiencies or failures, to eliminate costly and sometimes unnecessary periodic maintenance, and to significantly reduce the likelihood of equipment failures. CBM applications built on a real-time infrastructure detect potential failures more quickly and provide a rich environment for discovering cause-and-effect scenarios based on immediate availability of equipment operating data.

Condition-based monitoring focuses on increasing throughput by determining when to stop the equipment before it breaks down, thereby reducing unscheduled equipment downtime. Having real-time data enables initial indications of increasing standard deviation in a motor amperage or impeding problems with rotating equipment using lube chemistry analysis, thus increasing plant availability, reducing utility consumption, and stemming production losses. Condition-based monitoring coupled with digital CBM systems directly supports the elimination of waste, which is consistent with an operational excellence program (Pierce 2018).

Paul sees opportunities for coordination among all the teams with easy access to the equipment data. Internationally recognized standards, such as ISO 55001 and PAS 55, reinforce the idea that successful programs require participation from all organizational levels and alignment to core business missions. Delivering the information at the right time means people do not have to waste time searching for data. In fact, information flows enable increased visibility and awareness for continuous improvement. Collaboratively handling the notifications is imperative for a successful initiative.

Preventing Abnormal Situations   141

Peter summarized: "The goal is to keep the flow of materials with the necessary information to maintain planed production flow at all times." ProcIndustries can

- Gain visibility into all static process units and mobile equipment;
- Get raw material data from the receiving department;
- Deploy additional sensors to monitor speed and power;
- Adopt and streamline collaborative processes;
- Document best practices at the local and enterprise levels;
- Integrate process, equipment, and safety procedures with necessary information, data, methods, and analysis rules; and
- Build competence centers to maintain and transfer knowledge from older to younger workers.

Condition-Based Maintenance Using a Real-Time Data Infrastructure

Paul and Peter plan to build reusable equipment templates to simplify integration with their maintenance processes and streamline integration with other systems. The process of building equipment templates is the same as adding context to a process unit (see Chapter 3). There are additional ways to add context to data as required to solve particular problems, which are discussed later in this chapter.

Paul, the ProcIndustries' maintenance manager, expects their asset management program should see drastic improvements over their current methodology. The strategy they use in maintenance can be applied to other functions such as production, quality, accounting, environmental conditions, safety, engineering, and marketing. The system capability to classify the data online as well as to generate triggers and alerts is the foundation for all performance metrics. The combined analysis of all these factors results in a balanced, environmentally safe, profitable operation.

Figure 6.5 shows the gathering of data for several sources, which is enabled by the EIDI in real time. The strategy is as follows:

1. Data acquisition
2. Analysis assessment and notification
3. Maintenance program integration
4. Work order change management

Tracking equipment health within the infrastructure also allows for web-based analytical tools to be used to integrate maintenance status information for any requested time range. It is imperative to test the logic and workflow of a condition-based monitoring process that intends to automatically send transactions to CMMS. This can be done over a period or history of data to ensure that there is confidence established once the process is automated.

Condition Monitoring

Input	Analysis	Notifications	Action
Process parameter or diagnostic test	Compare limits, rate of change, correlation, APR, etc.	Alert to personnel or systems	Acknowledge or escalate, comment or adjust

FIGURE 6.5
CBM process workflows (data, assessment, maintenance policies, and work order).

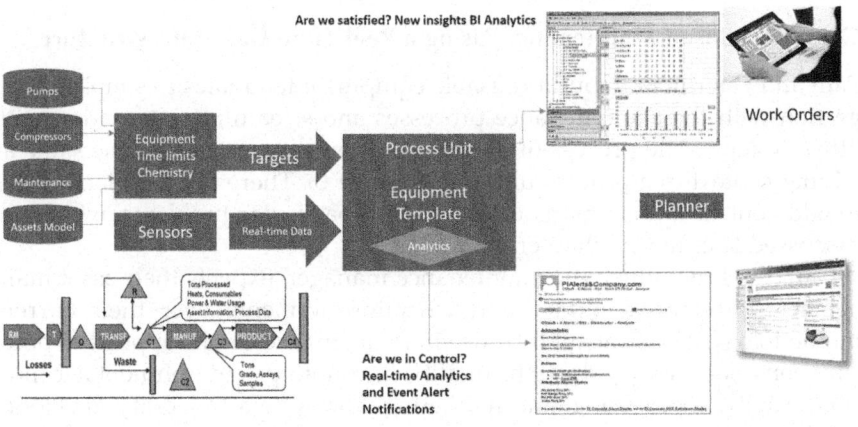

FIGURE 6.6
Condition-based maintenance strategy using EIDI data.

Figure 6.6 shows operational intelligence strategy schematics. The diamond represents the analytics portion of the automated procedure. The boxes represent the outcomes, which are various ways that people and systems benefit.

To get the ProcIndustries CBM initiative started, Paul and his team prepare sample templates for pumps and compressors. They include real-time collection of equipment sensor data and integration with targets or equipment parameters. They configure the object model and define proper analytics to trigger a notification to the appropriate people and their CMMS.

Assigning Context to the Equipment Template

Assets are represented as elements within the OSIsoft PI System Asset Framework (AF) templates and are organized for easy access and in support of analytics (e.g., rollups). This representation can include static data, access

Preventing Abnormal Situations

to streamed data, calculations to determine asset condition, and data from external systems (e.g., nameplate, date of last service, etc.).

The equipment template can be used as a sub-element of the process unit template discussed in Chapter 2. Assets can also be grouped by families. For example, an asset family can be rotating equipment such as centrifugal pumps, compressors, conveyors, and blowers and additional static equipment such as furnaces and heat exchangers. The asset object model is very flexible to accommodate user preferences.

CBM is an ongoing process. Teams begin by mapping some critical equipment and some critical attributes. Once assets, attributes, and their data streams are mapped in the asset object model, results are achieved every time somebody accesses data for CBM as well as other activities supporting operational intelligence.

Assigning Analytics to Detect Abnormal Conditions and Trigger Events

The analytics can range from simple to complex. For example, the system can identify periods of high vibration, periods of low pump efficiency, or a totalized run time during which a specific material is processed. Individual condition factors can be combined into a health score key performance indicator (KPI) for the asset.

Figure 6.7 shows an example of how maintenance rules are configured using the asset model. Triggering conditions for each rule are configured using dialogs for comparative, statistical, and equation-based tests. There are also Boolean dialogs for specifying logical "AND" and "OR" conditionals.

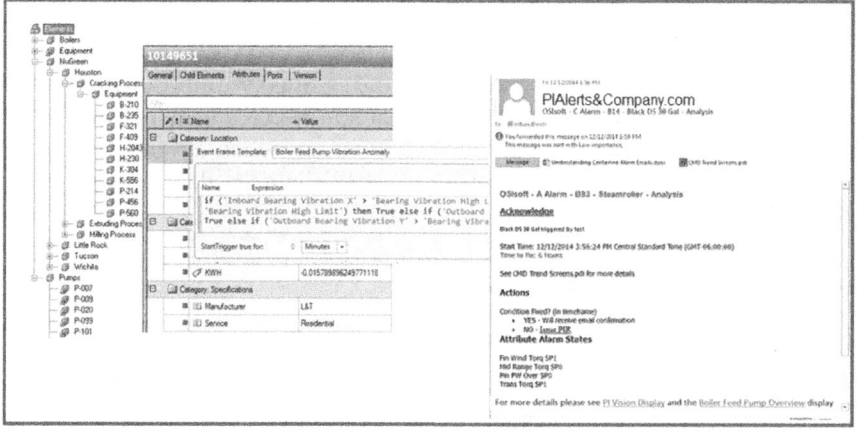

FIGURE 6.7
Configuration of analyses and notifications.

Generating Notifications from the Events

Analysis results and events can fire notifications to either alert an individual to a condition or update another enterprise system. For example, OSIsoft's PI Notifications software can send an email or text to an engineer indicating that a specific condition has occurred. It can also update the CMMS with a meter (operation) count.

Assigning the Event Templates for Analysis and Root Cause Determination

The event frame template simplifies the analysis of a repeatable event with a defined start and end time. CBM events can be layered to gain insight about how the assets have operated over long periods of times by comparing them to each other over these time segments. In Figure 6.6, an event frame is initiated by a trigger that could be a state change in process data. Process variable exceptions, such as a rate of change, crossing of a limit, or run time within a given period, can also initiate an event frame. Asset analytics use sophisticated calculations to define reference events based on process data. Because an event frame is always true after a defined start point, users can capture (or reference) data before the actual trigger. For example, when a unit or piece of equipment trips, users may want to see specific data for a period before this event to determine the root cause event or conditions.

Events capture all of the following information:

- Frames (or bookmarks) that define the event
- Configurable time period before the event trigger
- Referenced element (asset)
- Specific attribute data
- Other calculations, such as duration and cost

Event frames capture data as referenced and defined within the event frame template as well as provide placeholders for additional data to be added later. For example, users can add in cause code values when an engineer determines the cause of the event or programmatically adds data returned from the CMMS, such as the work order number.

Assigning Equipment Parameter Tables

Static characteristics about the equipment can be stored in the equipment parameter table if the real-time analytics in the real-time asset object model use them. As such, the analytics and rules for families of equipment types can be reused, simplifying the implementation and maintenance of these objects. These parameters might include the manufacturer, asset type, asset type parameter, size, nominal power, and associated data to trigger the events and notifications. This information is usually available in the CMMS environment.

Visualizing the information is easier when it is implemented using a digital plant template asset object model, which provides the attributes for each element to be queried by the visualization layer. This strategy simplifies the access and delivery of information for both real-time awareness and historical analysis. It also enables the historical data and events information to build predictive models or soft sensors from the data. *Soft sensors* are inferential estimators, drawing conclusions from operational data when hardware sensors are not available or when they have reliability issues. They provide an alternate way to have information when hardware sensor performance is sometimes inaccurate.

Paul provides a list of some common CBM scenarios:

- Pump lubrication preventive maintenance (PM) based on the actual motor run time, for example 2,000 hours
- Analyzer recalibration PM based on the measured drift exceeding 1%
- Filter change PM based on measured pressure differential across the filter exceeding allowable limits
- Heat-exchanger cleaning cycle PM based on calculated fouling factors
- Circuit analysis based on significant switching operations in a small amount of time (e.g., more than six operations in less than 24 hours)
- Transformer servicing based on dissolved gas analysis results over time that indicates insulation degradation
- Bearing swap-out based on vibration data of the motor shaft or motor current signature analysis

CBM can be started with limited process or asset information and does not necessarily require specialized diagnostic or condition-based monitoring equipment. For example, CBM can be used to monitor the amperage of the motor attached to pump. The analysis of the amperage in real time, including its variability within a minute, can be used to detect an anomaly for further inspection.

With data to define proper data parameters, ML can be used to generate patterns, resulting in real-time alerts when abnormal conditions occur. The generated models can be incorporated in the analytics or be executed as a sub-process (Cope and Chugh 2019). Figure 6.8 represents a typical trend for a piece of equipment (asset) showing predicted behavior and the measured real-time behavior of the asset. As soon as these values start to drift, a possible fault is detected and a notification is sent to the operations and maintenance departments for further analysis, aligning with the workflow described in Chapter 4.

EIDI Historical Trend Using Pattern Recognition Software

When Deviation Between Actual EIDI Data and Expected Values Derived From Pattern Recognition Software Sufficiently Diverge, A Real-Time EIDI Notification Can Be Sent

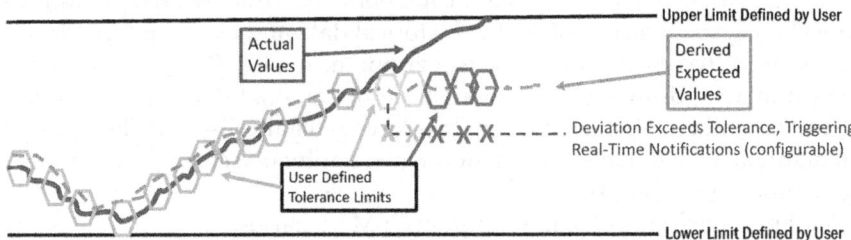

FIGURE 6.8
Advanced pattern recognition for equipment failure prevention.

Pump Asset Example

Pumps are one of the most common rotating equipment pieces in the world. They move fluids in the plants; they are the equivalent of the heart in the human body. However, the potential of pump CBM as a source of improved productivity and reduced costs is often overlooked. Because companies utilize numerous types of pumps, creating a standard strategy to proactively perform maintenance can be perceived as a daunting task. A standardized approach, however, leads to a variety of improvements, including the following:

- Improved efficiency (operating close to the best efficiency point [BEP] of the pump)
- Prolonged asset life (through more effective usage patterns)
- Improved capital expense programs (by better equipment sizing)
- Lower energy costs (by optimizing energy management)
- Maintenance improvements (by performing the right maintenance at the right time, moving from unplanned to planned activities)

The ProcIndustries team discusses an approach to leveraging the EIDI to create a standard, scalable monitoring method for the organization's centrifugal pump assets. Alex Moretti, the process engineer, explains that one of the most frequent decisions that a process operating engineer is asked to make is whether a pump needs to be repaired. Troubleshooting pumping difficulties is a job that operating engineers face every week. A major part of a process unit's mechanical budget is allocated to pump maintenance.

Pump problems are caused by several things, with cavitation one of the most common problems. A cavitating pump makes a rattling sound. To avoid cavitation, a pump needs to have a net positive suction head (NPSH).

Preventing Abnormal Situations

This means that the level of the liquid in the tank has to be high enough to ensure that the fluid is above the bubble point pressure at the impeller. Figure 6.9 shows a typical pump curve and the system resistance curve from the piping system. Monitoring the state of the pump is very important to avoid process and equipment problems. Many pumps are sent to the maintenance shop because of excess vibration. Often there is nothing wrong with the pump. The problem may be a low liquid level in the sump from which the pump is taking suction. The level controller or the suction line restriction may be the cause of this problem. To monitor the efficiency of the operating pump, motor, and mechanical variables (such as vibration and lubrication), aggregating the process data together with the pump mechanical measurements makes sense. Plotting the head measurement versus flow rate is key in identifying anomalies in the pump. This is also valid for other rotating equipment such as compressors. However, compressors are much more sophisticated than centrifugal pumps.

Figure 6.9 shows the head capacity and piping system resistance curves. The total head (*middle downward arc*) shows the head capacity curve of a centrifugal pump. The pump can operate only along its head capacity curve

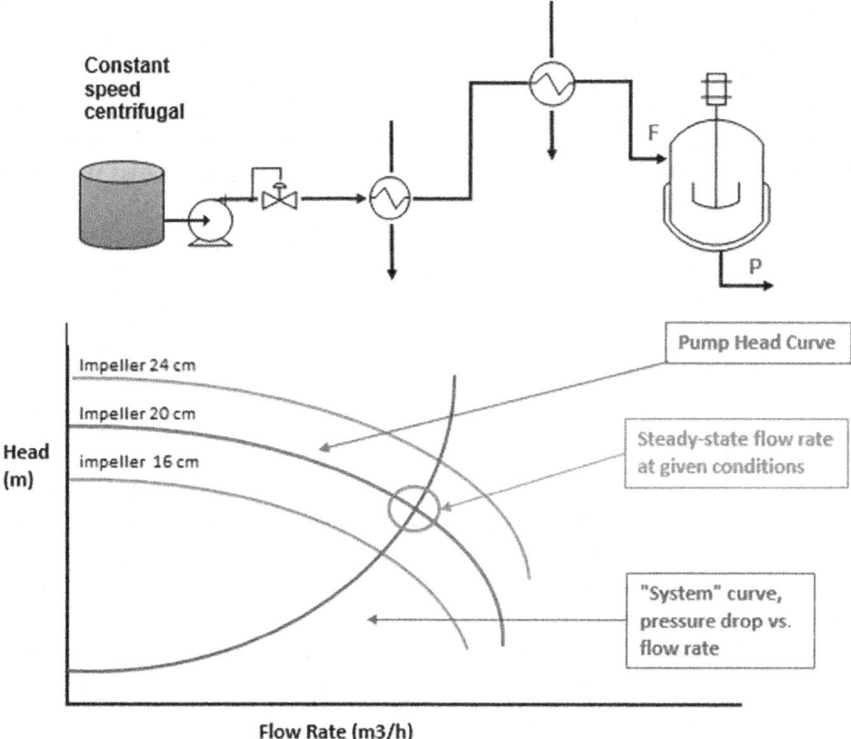

FIGURE 6.9
Typical centrifugal pump system.

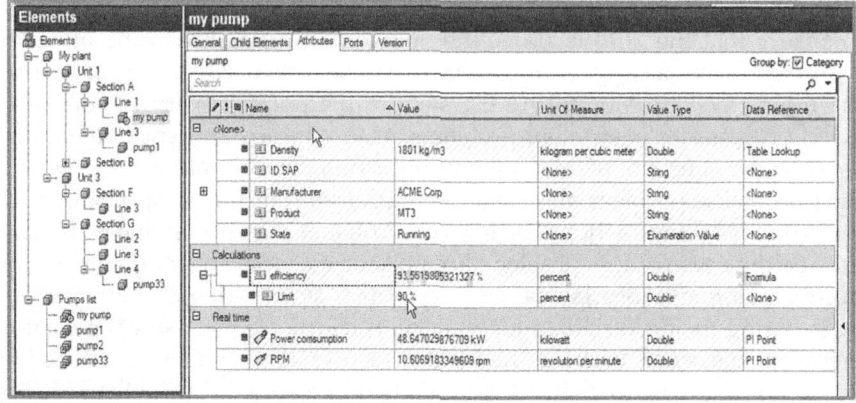

FIGURE 6.10
Pump operational efficiency estimate for performance analysis.

with the corresponding capacities and total heads. Only one resistance curve exists for a given pipeline. The gap between the head capacity curve and the system resistance curve is available for throttling resistance (control valve delta P). Manufacturers typically provide a performance characteristics diagram for each pump. The pump curves enable the selection of the correct impellers and pump speed required for the hydraulic process. Avoiding cavitation of the pumps because of a bad suction head design or bad operating conditions is imperative.

Figure 6.10 shows the attributes for a centrifugal pump with power consumption (head) versus revolutions per minute as a way to estimate the current operational state of the pump based on the design parameters. An operational alert can be defined using notifications to alert operations if the pump is not within the operational window for normal operation.

Pump Monitoring and Analysis

A principle challenge in creating an effective maintenance system for pumps is producing a systematic framework to link static pump properties, high-fidelity operational data, and operational context. Examples of attributes, data streams, and events that should be brought together so that users can see asset information in context are shown in Table 6.1.

Having this information in the EIDI means that pump data can be organized according to asset attributes and topology (Figure 6.11). The asset object model creates a template for a basic pump that includes available information that applies to the majority of pumps, such as nameplate data, maintenance data, date installed, criticality to the process, next scheduled maintenance date, and so on. Once templates are populated with data

TABLE 6.1
Pump Characteristics and Variables

Manufacturer	Best Efficiency Point (BEP) (Total Dynamic Head [TDH], Efficiency, 80%, 110%)
Type	Pump on/off pump run hours
Size	Pump start/stop number and frequency
Horsepower	Motor temperature
Available net positive suction head (NPSH), a	Flow rates and pressures (in, out)
Required NPSH, r	Vibration monitoring points
Static head	Bearing temperatures
Friction loss	Motor current signatures
EQ number	Motor power, kW
Process	Visual inspection results

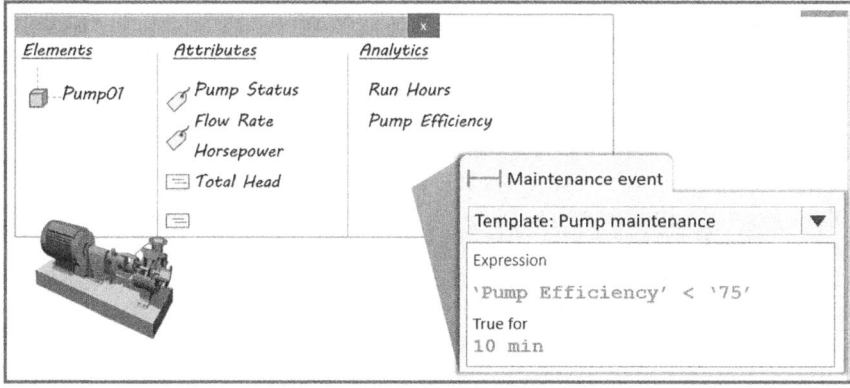

FIGURE 6.11
Simplified pump efficiency evaluation to generate an event.

streams, calculations, and static data values, users have access to a single, consistent method to shape and prioritize pump maintenance. CBM implementations are more costly to manage through manual inspections.

For example, users can track run hours of the pump or track basic conditions related to the operation and maintenance of the pump. When pump issues or failures occur, historical data can be analyzed to develop condition indicators and link root causes to vendor, use conditions, or external conditions. Once this first phase is complete, the team can consider implementing a more robust strategy of incorporating alarm data and integrating the EIDI with a CMMS system (to drive a true CBM solution). For example, if the number of pump run hours is approaching the vendor-specified limit for maintenance or calibration, operators can use EIDI to define a trigger that creates a work order in the local CMMS system. When this work order is complete, a message can be returned to reset the run-hours counter on the asset. Now that both systems are configured for operations, maintenance, engineering, and production planning, the various teams will be able to collaborate and support each other in numerous ways.

Simply deploying condition-based monitoring on plant pumps can lead to improvements in operation and maintenance and is a first step to true CBM. Condition-based monitoring can be accomplished using operational data and condition-specific data and eventually lead to information that serves as the basis for other predictive techniques, such as APR, which can be particularly useful in pump maintenance.

Visualizing Pump Actionable Output

A key part of an effective PM solution is to develop accurate condition indicators and methods to display actionable asset condition information. The EIDI system provides visualization options that can be reused, shared throughout the enterprise, and easily modified as operations and data needs evolve. Operators can use real-time displays to monitor asset conditions within any maintenance strategy (Figure 6.12). Users can create standard visuals and KPIs of pump data for knowledge workers in multiple roles using EIDI visualization tools (OSIsoft's PI Vision).

When organizations use the EIDI asset object model to structure asset data, operators, maintenance personnel, or engineers can use asset-relative displays that toggle between different pump assets associated with the same template. Over time, users can determine how one vendor's pumps perform relative to other vendors' pumps or comparative maintenance costs. Users do

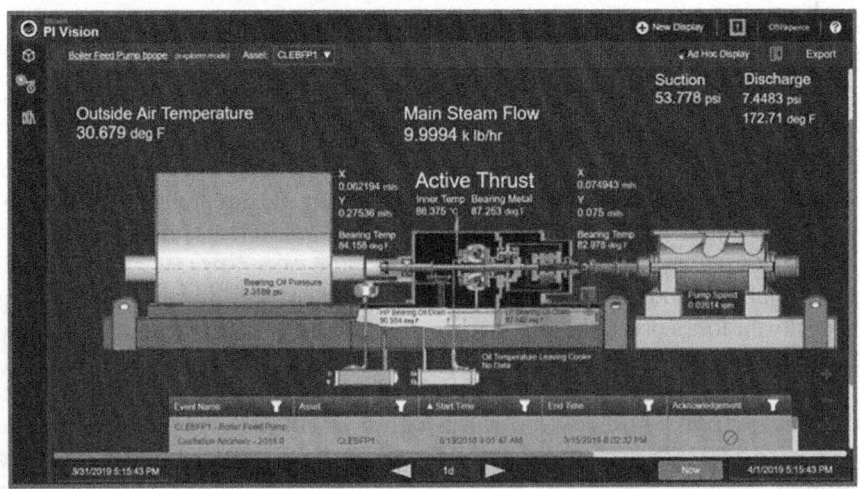

FIGURE 6.12
Sample condition-based monitoring display for field checks.

Preventing Abnormal Situations

Sample Run Hours Display

Report - Critical Motors - Run Hours   Last Update: 3-12-2016

Equipment	Daily Run Hours	Lifetime Run Hours	During Last Period	Period	Since Last Service	Last Service	Next Service
Agitator 1204	4.51	7,975	0	3mo	387	1/10/2016	11/10/2016
Agitator 1205	23.79	10,119	2,154	3mo	409	2/23/2016	10/3/2016
Agitator 1304	23.49	9,908	2,118	3mo	697	2/11/2016	12/13/2016
Agitator 1305	23.49	9,908	2,118	3mo	697	2/11/2016	12/1/2016
Fan 5163	19.71	8,554	1,174	3mo	2,664	10/1/2015	5/1/2016
Fan 5164	23.97	9,292	2,022	3mo	3,566	10/2/2015	5/2/2016
Fan 8144	14.44	9,839	2,112	3mo	3,635	10/5/2015	5/5/2016
Pump 3809	15.16	8,587	1,949	3mo	3,218	10/10/2015	5/10/2016
Pump 3810	23.97	9,618	2,079	3mo	3,837	9/23/2015	7/1/2016

FIGURE 6.13
Equipment totalized run time with service history information.

not have to create individual graphics or calculations for each pump. Using standard asset templates simplifies the creation and maintenance of the data management system.

Maintenance and engineering personnel can drill down into archived condition-based monitoring data for post hoc analyses. Engineering can conduct root cause analyses, develop more precise indicators to optimize CBM solutions, or create analytics for equipment replacement evaluation. In some cases, maintenance may need to access this information to evaluate what maintenance to perform and when to perform it.

Another example report showing totalized run times and service history is shown in Figure 6.13.

Integration with Other Systems

Asset data can be integrated into CMMS and other enterprise applications in a number of ways. There are two basic scenarios: *push* or *pull*. Push sends data to people and systems using any number of delivery channels, for example, by sending emails, texts, or invoking a web service. The pull model is enabled by EIDI data access technologies where enterprise applications can acquire data from the EIDI using any number of technologies, including the most popular open standards methods such as open database connectivity (ODBC) provided via the EIDI SQL Explorer. (See Chapter 10, which focuses on the EIDI integrator as a tool that can be used to simplify

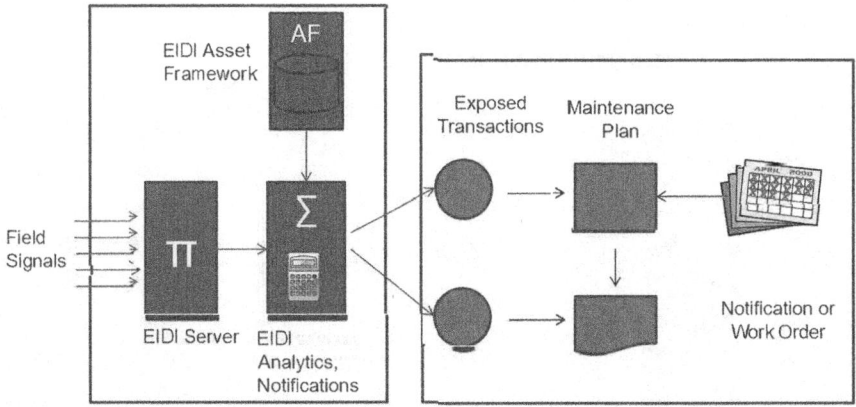

FIGURE 6.14
Example of EIDI real-time notifications and work orders.

the integration with business systems and advanced data analytics tools that are available in the market.) Pierce (2018) provides several scenarios for CMMS integration.

Figure 6.14 shows how the EIDI interacts with a maintenance application. The generated triggers are compared with the maintenance plan logic to create a work order for repair of the equipment. This work order triggers a process change management system to assure that all process safety guidelines are properly followed. As ProcIndustries connect to their data using the digital plant template, the operating and maintenance teams are brought together. They all share the same information with the capability of defining the proper triggers to generate alerts for further analysis by the maintenance and engineering teams. As such, they will be able to make an informed recommendation as to when to execute a work order.

Paul summarized: "The team will be able to use data mining tools to do a more detailed investigation. So whether you call it 'big data' or simply connecting people, I believe that bringing data from silos connects people and encourages them to use their knowledge for process improvements and innovation."

Implications of Improving Asset Availability

Monica shares that from a production perspective, avoiding costly repairs and downtimes will improve quality management and align with the production schedule. She adds that each production team has to provide the best internal product on time for the next operation to have the least possible disturbance. In the past, ProcIndustries handled large inventories of products that needed reprocessing. Monica points out that these events will be minimized as well.

Reducing waste product (also known as off product) inventories increases potential plant throughput and overall plant production. By abstracting the process unit, time constants, and inventories for all processing areas, users can visualize all related information using a Gantt chart to instantly check daily schedule performance.

Peter added, "Big data is a relatively new technology for us. Now we can capture and integrate real-time data and events in our business."

"Health, safety and environmental issues are always going to be the first priority in the process industries," says Raj Singh, the process safety manager. "Critical situations always arise when we have downtimes or unexpected situations. Using CBM will help us avoid safety and environmental issues. Although we can monitor minor losses, which are very hard to solve, we still have a long way to go. Now that we have the measurements of all the losses clearly defined, just having these metrics will change our corporate culture."

Trouble time was a hidden loss that was not possible to detect without having a way to trigger events and be able to automatically aggregate production consumables data. Learning from the collected data and operational events will allow for better planning and improved production campaigns. Monica is very proud of what they have achieved in such a short time: "Capturing these hidden losses as key metrics opens a new way to improve our processes. This is really the integration of operations and business."

In Chapter 2, the strategies for continuous improvement enabled by a data infrastructure system were presented. In this specific case, the operational and equipment data is processed by real-time analytics to generate events and the alerts for finding the root cause. Once the root cause is known, the maintenance workflow can begin to take the best action to keep the plant running, or to schedule repairs aligned with the plant's production schedule.

Shared analysis is readily available, offering a view of all information related to an exception raised by the CBM system. Web-based portal technologies are best suited for establishing collaborative environments because they allow for easy integration of real-time, relational, and unstructured (documents) information. Portal technology eliminates the need for client software licenses or training when providing views of information for decision-making.

The benefits of a well-implemented CBM program extend outside of protecting corporate investment in asset portfolios. When real-time asset data provide visibility into asset condition, maintenance schedules and costs can be planned and spent correctly. (Will that compressor make it until the next outage?) Common failures or issues that occur across units or within fleets can be identified. (Why is maintenance more expensive on a specific vendor's equipment compared to other vendors?) Just by creating visibility into asset condition indicators, data can help prevent catastrophic failures. It typically only takes a few big saves to pay for a complete CBM implementation.

From what Peter and Monica have explained, Paul knows that real-time, online health monitoring for pumps and other rotating equipment can be constructed using off-the-shelf EIDI components. The key business strategy is to eliminate unscheduled downtimes to zero. He summarizes the next steps for the EIDIs:

1. The interfaces collect the required data from the pump motors, bearings, flowmeters, and so forth. These typically come from PLC, DCS, or supervisory control and data acquisition (SCADA) interfaces and IIoT devices.
2. The server routes data to all analysis routines and visualization clients in real time. All aggregated data, both sensor and calculated data, are kept in the EIDI historical data archive.
3. The asset object model organizes pump data according to asset attribute, elements, and topology to create a standardized method of assessing pump health.
4. The asset analytic is used to calculate real-time condition indicator values.
5. The statistical quality control (SQC) system compares the computed values against statistical norms.
6. Contextualized extracted EIDI data enables offline tools, such as advanced predictive analytics and AI. These tools do not require real time execution.
7. The EIDI visualization tools (OSIsoft PI Vision, PI DataLink, and other client tools) display real-time and historical information in visual and graphical formats, enabling role-based dashboards, and deep content displays.

Operational Performance Management

Enhancing Process Control Performance Monitoring

Another critical maintenance issue is the process control and sensors asset health. As we saw in Chapter 4, the critical operational hidden losses when the units are in trouble are abundant. These hidden losses are related to operators not fully understanding the advanced controls and setting them to manual mode. These issues can be related to bad controller tuning parameters and/or having the final control elements not capable of performing the process control duties. As equipment ages, all assets degrade over time and the process controls do as well.

Control-loop monitoring is another use case candidate enabled by having an EIDI. As Peter and Monica demonstrated, hidden losses are key detriments to profitability. These losses are often created when the operator manually controls units designed for advanced process strategies. As such, process control availability, sensors, final control elements, process flow diagrams, and tuning need to be constantly monitored and supported by the central control team. The control strategies should be optimized, but they also need to be integrated, and team members need to make sure the strategies do not adversely affect safety or employee health or create environmental problems.

This strategy focuses on process control-loop monitoring, instrumentation calibration, and process stabilization and optimization. The goal is to monitor the key metrics for the unit and unify the process control loops associated with the unit optimization strategy. Control loops require continuous monitoring and tuning to maintain their performance and usability. There is always degradation of the sensors, final control elements, or process changes in the equipment that need to be accounted for (Beal 2016a, 2016b).

Maintaining all the control loops is a challenge in all process plants. Being able to track the control and manipulated variables against the process disturbances allows for improved product quality and overall management of the process on a day-to-day basis (Smith 2016a, 2016b, 2016c; Ruel 2010a, 2010b).

At ProcIndustries, control-loop performance monitoring not only helps identify faulty control valves, it also provides an indication of properly functioning valves that may have been targeted for maintenance through a PM program. The cost avoidance of not performing maintenance can be significant (Van Scholkwyk 2012).

There are several control performance goals that should be taken into consideration:

- Controlled-variable performance: standard deviation, variability, percent of time inside limits, percent of time not saturated, noise level, oscillation index
- Model error: variability, integral moving average
- Control-loop performance: percent of time in highest mode, percent of time in service, percent of time in normal range, oscillation index, maximum error
- Manipulated-variable behavior: percent of time not saturated, error due to manipulated device, sticky valves, valve travel, valve reversals

Sensor monitoring is intended to validate data and detect failure so as to reduce risk of equipment damage, and improve overall production performance and availability. The primary benefits of intelligent field devices are improved process control and optimization as well as the ability to better manage the health and life cycle of the devices.

Chuck said, "We can build an asset data model for the controller and create a standard cookie-cutter object model for all controllers in the refinery." Peter seconded Chuck's idea: "The asset data object model for a controller can be added for each of the process units. We have object models for pumps, compressors, exchangers, and now for process controllers."

A key boiler metric is steam production (Table 6.2). This key metric is associated with the performance of the feed water controller and the drum level controls. The same analysis used for the boiler can be extended to all process control loops in the refinery process units.

Displaying the operating variables for all controllers and checking the operating window limits help detect if the process requires further attention. As such, the team can analyze operating data to select which process controllers require scheduled maintenance. For example, if the percent output is near 90%, this means the controller actuator might be maxed out. If a controller is operating manually, this means that the operator does not trust the control tuning and has decided to place it in manual. Using Chuck's data format allows the team to improve operability of the process units. The basic attributes required are the same as shown for the boiler drum.

Chuck explains that he had been studying critical control-loop data in his spreadsheet. He now has moved these calculations to the EIDI, placing them into the asset data model real-time analytics. He has defined an event frame

TABLE 6.2

Simplified Control Monitor of a Boiler Drum

Key variable	Steam production					
	Controller loop	Percent on	Output, %	Stability	Average absolute error	Error, %
	Feed water flow	Yes	50	5	3	10
	Drum level	Yes	65	3	2	9
Other units						

for each of the process units and associated the list of process controllers with it. His new asset data model for control-loop monitoring includes the following:

- Controller loop description
- Process variable
- Disturbance variables
- Setpoint variable
- Output variable
- Controller mode (manual, cascade, off, automatic, etc.)
- Error between the process variable and the setpoint
- Event generator to catch the controller mode time intervals
- Minimum, maximum, and average standard deviation associated variables

New software-based tuning capabilities and integration with real-time process analytic tools require not only process control knowledge but IT skills for data mining, ML, and statistical analysis of business and refinery operations. In addition, all process control efforts must be integrated into and compliant with refinery environmental and safety practices.

The tuning basics of self-regulating processes are very clearly understood, while advanced process controls are evolving toward overall process optimization. Good summaries of process controls basics are outlined by Beal (2016a, 2016b) and Smith (2016a, 2016b, 2016c).

What You Should Take Away

Production companies are now capable of keeping their critical assets in peak or near-peak running conditions and increase scheduled asset availability to approach 100% uptime, using the EIDI effectively, through methods described in this chapter. Companies will be able to minimize unscheduled downtimes, avoid lost production, and significantly reduce parts and labor costs for unscheduled equipment repairs. Once a company has implemented a reliable data-driven CBM program, predictive analytic tools can take this a step further by determining when assets are predicted to fail. This knowledge provides companies the ability to more effectively budget for new equipment, schedule site outages, and conduct repairs.

This chapter also provides an evolution of plant controls and how companies are using more business information as part of their control strategy. As we peer into the future, it is likely that the next generation of plant/refinery control systems include software components such as ML, statistical modeling, and predictive features.

Because experts can be located anywhere, the maintenance notification system is capable of sending messages to a variety of destinations such as email or phone texts. Notifications provide the recipient with the ability to acknowledge the alert directly from the email message or remote device. All notification and acknowledgment activity is logged by the notification server and is made visible to everyone involved in the maintenance process so that they can effectively cooperate to solve the problem. Escalation functionality is used when a notification is sent and the EIDI does not receive an acknowledgment within a prescribed time (Bascur et al. 2016).

In addition, the team has followed best practices by incorporating maintenance of their sensors, controllers, and associated equipment, as these also degrade and require special attention to avoid costly hidden losses. Figure 6.15 represents an advanced asset management system deployment, with automated inputs, online analytics, and connectivity to work management systems and asset data bases. Often there is a dedicated group of subject matter experts within the enterprise who initially get this data to interpret whether issues are urgent, or can be scheduled for the next maintenance event. This approach allows production and operations people to focus on meeting production targets without having to decide whether to shut down equipment during product runs. These multiple groups coordinate to determine the best course of action when situations arise.

Integrated Asset Health Management System

- Continuous improvement is the goal.
- Continue to tie the processes and systems together.
- It's a journey, not a destination, as new analytics always emerge.

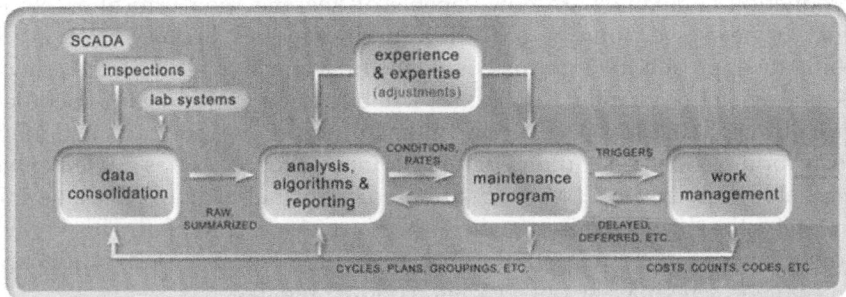

FIGURE 6.15
Integrated asset health management system approach.

For a complete guidebook on asset management, please click the link to OSIsoft's condition-based maintenance best practices document at www.osisoft.com/predictivemaintenance.

References

Bascur, O.A., and Aroqui, R. 2014. "Drastic reductions in energy and water consumption in large iron and steel metallurgical complexes." In *AISTech Proceedings*. Pittsburgh, PA: AIST.

Bascur, O.A., and Halhead, M. 2013. "Energy effectiveness and sustainability management at Anglo American Platinum." In *Proceedings Copper 2013 International Copper Conference*, vol. 1, ed. C. Moscoso, J. Rosales, and A. Vio. Santiago: Chilean Institute of Mining Engineers (IIMCH). pp. 415–423.

Bascur, O.A., Halhead, M., Garrigues, L., and Jarvis, M. 2016. "Mineral processing plant asset and energy optimization: The calming cloud over operations." In the *International Mineral Processing Congress (IMPC XXVIII) Proceedings*. Quebec City: IMPC.

Bascur, O.A., and Soudek, A. 2019. "Grinding and flotation optimization using operational intelligence." *Mining, Metallurgy & Exploration* 36(1):139–149.

Beal, J. 2016a. "Loop tuning basics: Integrating processes." *Intech Magazine* (March/April): 48–51.

Beal, J. 2016b. "Loop tuning basics: Self-regulating processes." *Intech Magazine* (May/June): 48–51.

Cope, K., and Chugh, S. 2019. "Predictive analytics for reliable high power performance, Ecolab Nalco Water." Presented at PI World, San Francisco, CA.

Dewald, D., and Santhebennur, A. 2019. "Barrick Gold Cortez process turns to industrial IoT to maximize uptime of its critical assets." Presented at 2019 PI World, Mining, Materials, and Supply Chain, San Francisco. www.osisoft.com/Presentations/Barrick-Gold-Cortez-Process-Turns-to-Industrial-IoT-to-Maximize-Uptime-of-Its-Critical-Assets/.

Pierce, K. 2018. *A Guidebook to Implementing Condition-Based Maintenance (CBM) Using Real-time Data*. San Leandro, CA: OSIsoft.

Ruel, M. 2010a. "Closed loop tuning vs. open loop tuning: Tuning all your loops while the process is running is now possible." bba.ca/publications.

Ruel, M. 2010b. "Control system performance assessment—best practices." In *Proceedings of the Second International Congress of Automation in the Mining Industry (Automining 2010)*. Gecamin, Santiago. p. 54.

Smith, C.L. 2016a. "PID explained for process engineers: Part 1: The basic control equation." *CEP Magazine* (January): 37–55.

Smith, C.L. 2016b. "PID explained for process engineers: Part 2: Tuning coefficients." *CEP Magazine* (February): 27–33.

Smith, C.L. 2016c. "PID explained for process engineers: Part 3: Features and options." *CEP Magazine* (March): 58–51.

Steyn, J., Bascur, O.A., and Gorain, B. 2018. "Metallurgy analytics: Transforming plant data into actionable insights." *Mining Engineering* 70(9):18–29.

Van Scholkwyk, T. 2012. "PI applications in AngloAmerican Platinum, control loop monitoring and reporting." Presented at the OSIsoft Regional Seminar, Johannesburg.

Additional Reading

Alyeska Pipeline: Compressed Air Challenge. n.d. "Achieving reliability-centered maintenance and diagnostics with the PI System 2012: Retrofitting compression equipment and interfacing systems for CBM improving reliability with Caterpillar condition monitoring services." www.compressedairchallenge.org.

Blanchard, B.S., and Verma, D.C. 1995. *Maintainability: A Key to Effective Serviceability and Maintenance Management*, 2nd rev. ed. New York: Wiley-Interscience.

Djuric, V. 2002. "How the PI played a key role in achieving maximum equipment reliability at Dofasco." Presented at the OSIsoft Users Conference, Monterrey, CA.

Dudzic, M. 1998. "The use of multivariable statistical technologies." Presented at the AISE Specialty Conferences, MIT, Boston, MA.

Efficiency New Brunswick. 2010. *Energy Management Information Systems: Planning Manual and Tool*. Ottawa: Natural Resources Canada.

Ferrari, A., and Russo, M. 2016. *Introducing Microsoft Power BI*. Redmond, WA: Microsoft Press.

Gopal Krishnan, G., and Hertler, C. 2018. "Fit for purpose: Layers of analytics using the PI System—AF, MATLAB, Machine Learning." Presented at the PI World 2018 Lab, OSIsoft, San Leandro. https://pisquare.osisoft.com/people/gopal/blog/2018/07/11/fit-for-purpose-layers-of-analytics-using-the-pi-system-af-matlab-machine-learning.

Harjunkoski, I., Scholtz, E., and Feng, X. 2014. "Smart grid: Industry meets the smart grid." *Chemical Engineering Process* (August): 45–49.

Johnson, D. 2015. "Pervasive sensing save money and prevents downtime." *Intech Magazine* (July/August): 66.

Kennedy, J.P. 2018. "The driving force of industry." *Scale and Scope* (September/October):34–37. www.isa.org/intech/20181005.

Kennedy, J.P. 2019. "Retrospective: Then and now—looking toward the future." *Hydrocarbon Processing* (January). www.hydrocarbonprocessing.com/magazine/2019/january-2019/columns/retrospective-then-and-now-looking-toward-the-future.

Kern, R. 1972. "How discharge piping affects pump performance." *Hydrocarbon Processing* (March): 89–93.

Liebermann, N. 2009. *Troubleshooting Process Operations*, 4th ed. Tulsa, OK: Penn Well.

MacGregor, J.F., and Kourti, T. 1998. "Multivariable statistical treatment of historical data for productivity and quality improvements." Presented at FOCAPO 98, Snowbird, Salt Lake City.

Moubray, J. 2001. *Reliability Centered Maintenance*. New York: Industrial Press.

Plourde, M., Bascur, O.A., Paquet, S., and Gervais, D. 2017. "Digital innovation in modern engineering and operational excellence." Presented at the 2017 SME Annual Conference and Expo, Denver, February 19–22.

Rossiter, A.P., and Davis, J.L. Jr. 2014. "Identify process improvements for energy efficiency." *Chemical Engineering Progress* (September): 53–58.

Ruel, M. 2014a. "Advanced control decision tree." In *Proceedings 46th Canadian Mineral Processors*. Westmount, QC: Canadian Institute of Mining, Metallurgy and Petroleum.

Ruel, M. 2014b. "Successful methodology to select advanced control approach." Presented at the Process Control and Safety Symposium, International Society of Automation, Houston, October 6–9.

Scheihing, P. 2014. "Save energy through the superior energy performance program." *Chemical Engineering Progress* (September): 48–51.

Tomlingson, P.D. 2009. "Reliability-centered maintenance improve equipment health." *Mining Engineering*, pp. 41–49.

Weber, C., and Yeung, T. 2010. "Energy management raising profitability, reducing CO_2 emissions." *Petroleum Refining Quarterly* 4:87.

7

Energy Management and Operational Improvements

If at first you don't succeed, take the tax loss.

Kirk Kirkpatrick

Chapter Overview

The South Texas refinery has deployed their enterprise industrial data infrastructure (EIDI) system with a few nascent results. Refinery equipment is being proactively monitored and the units are running more smoothly, with a reduced number of unscheduled downtimes. Engineers and operation teams are better able to monitor their processes, resulting in less variability. The team now turns its focus to reducing energy costs and improving their analytic abilities. In this chapter, we will investigate how ProcIndustries can use their EIDI technology investment to benchmark and reduce unnecessary energy costs. We will also see how Peter Argus and his team strive to make refinery operations more resilient, so that it can survive another disaster like the transformer lighting strike, which triggered refinery outages. This incident ultimately spawned the strategy of EIDI deployment.

Revisiting Energy Consumption

One of the largest costs for a process or manufacturing company is their consumption of utilities: steam, water, and especially power. For example, many manufacturing production companies treat energy costs as fixed costs, with energy (power, steam, or water) consumed and the monthly utility bill paid. With consolidated real-time and historical data, a company can investigate when and why they are not optimally consuming energy, which assets devour the most energy, and which assets ingest the most energy relative to expected consumption. This type of analysis will help ProcIndustries avoid paying stiff penalties for overusing energy during peak summer afternoons at the South Texas refinery. The team can also allocate utility costs by process, by unit, by product, and other metrics.

Ron Erickson, ProcIndustries' energy and utilities manager, who has been working with the digital transformation team, met with Peter Argus and the refinery's transformation team. Ron saw a new way of reengineering energy management for ProcIndustries. "First, we must not look at this as a one-time event, like we've done in previous energy audits, when management only wanted to replace a few faulty pieces of equipment and change the lighting system. I think we need to go deeper, now that we have the data to do so." He surmised the enterprise industrial data infrastructure (EIDI) will enable the following:

1. Establish a baseline level of energy consumption. The EIDI will monitor current energy usage for a month under normal refinery operating conditions.
2. Investigate ways to reduce the refinery's monthly energy bill by making sure assets run optimally (refer to Chapter 6 on ProcIndustries' condition-based maintenance initiative) and by reducing controllable process losses that gobble energy.
3. Translate usage to actual energy costs per unit by adding EIDI online calculations.
4. Provide real-time visibility for plant operations teams to identify when energy usage starts to deviate from desired target ranges or historical norms. In those instances, they can take corrective actions or notify others to investigate.
5. Improve refinery resiliency by using the EIDI to notify operations of cascading consequences from calamitous events, such as the lightning strike the previous year.
6. Use the EIDI to continuously monitor and reduce the refinery's energy consumption through intelligent data analysis. As people better learn how to use the EIDI, they will naturally go deeper to find correlations.

Ron said, "Energy affects refinery margins," and shared the following statistics: "For U.S. refineries, energy accounts for the largest share of operating costs at more than 40%. Maintenance is about 20% to 25%, personnel is 10% to 20%, and 15% to 20% for everything else. And because water is used for cooling, it is also an energy component. In addition, unscheduled equipment downtime reduces our annual production by about 5%. An energy-efficient refining program can be achieved with results in the order of a 4% to 6% payback per year. We can also use EIDI to track carbon dioxide, sulfur and nitrogen oxides, and volatile organic compound (VOC) emissions." Ron concluded, "If we accomplish those goals, we can improve profitability and become a better neighbor in the community by meeting our corporate sustainability goals."

International Energy Standards

The International Organization for Standardization (ISO) has established the ISO 50001 standard in order to significantly reduce energy consumption by commercial and manufacturing entities around the world. This standard can influence up to 60% of the world's energy use by enabling organizations to establish systems and processes that are necessary to improve energy performance, efficiency, use, and consumption. The ultimate goal is reducing greenhouse gas emissions, environmental impact, and manufacturing energy consumption. Their ISO 50001 standard provides guidelines regarding energy management.

ISO 50001 provides a framework of requirements for organizations to

- Develop a policy for more efficient use of energy,
- Fix targets and objectives to meet the policy,
- Use data to better understand and make decisions about energy use,
- Measure the results,
- Review how well the policy works, and
- Continually improve energy management.

When a company embarks on an ISO 50001 journey, its success depends on the commitment of all levels and functions within the organization, particularly upper management.

Ron investigated several approaches and decided to follow the ISO 50001 methodology as well as energy management guidelines recommended by Natural Resources Canada (2019; see link National Resources Canada Energy Mgmt.) using the EIDI as the system of record. Figure 7.1 presents a diagram of continuous improvement, integration, communication, and optimized performance.

Develop Clear Business Objectives

The purpose of an energy management information system is to reduce energy consumption while optimizing production. A series of process, equipment, laboratory, weather, and energy data is collected from several systems. A data infrastructure management system aggregates these types of data and simplifies data analysis into several functions and targets for overall process effectiveness.

Using the EIDI as the real-time data system of record can assist in reaching these objectives. Perhaps the most beneficial EIDI advantage is that the same

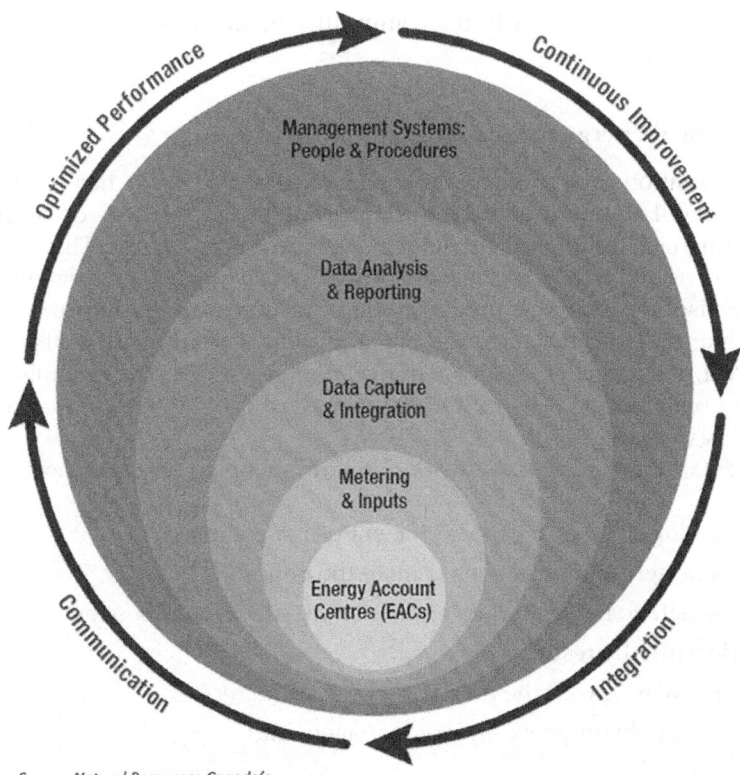

Source: Natural Resources Canada's
"Energy Management Information Systems Planning Manual and Tool"

FIGURE 7.1
Natural Resource Canada's energy management methodology. (From Operational Energy Management Information System [EMIS] from Energy Management Information Systems, Planning Manual and Tool, *Natural Resources Canada*, 2010. Reproduced with the permission of the Department of Natural Resources, 2019.)

data can be used and reused for many different applications and use cases, such as energy management. Capturing all process, production, and equipment data at their original resolution facilitates many types of analyses, including the correlation of energy usage to refinery or plant conditions, such as

- Identifying which assets are not running in optimal condition and determining if they contribute to excess power consumption;
- Analyzing which plant processes can be tweaked to use less power, steam, or water;
- Determining if there is energy use variability among plant operator crews; and
- Coupling equipment production modes, such as idle or in downtime, to energy usage.

Energy Management and Operational Improvements 167

The plan is to continuously capture the energy data (electricity, fuel, and steam flows) for each unit and classify the information in the context of process steps and production runs, such as when the refinery is making different products. This will also enable ProcIndustries to visualize current losses when the process is not running on target. Ron proposed integrating production, energy usage, emissions, and process availability to monitor the cost and availability of electricity in real time, with notifications generated before issues become costly problems.

Several early adopters have achieved considerable savings in their energy usage while increasing the availability of the assets using the continuous improvement strategy presented in Chapter 2.

The Plan

After Peter Argus' transformation team has several discussions with Ron Erickson, they make a list of the steps they will take to analyze and reduce the refinery's energy demand, or consumption:

1. **Real-time energy monitoring.** The team will monitor energy usage for one month during seasonal weather and during typical production rates. Once refinery energy usage is baselined through EIDI inline calculations and historical analysis, EIDI run-time analytics will monitor and check that consumption does not deviate from or exceed historical norms for a unit, process, product, and so on. If it does, it will generate real-time alerts via the EIDI to forewarn operations personnel that the process is using excess energy, and personnel should take corrective actions.
2. **Energy event management.** Detect and analyze process changes that cause consumption to exceed forecasts. This is done via online forensic analysis and by extracting process data for input into advanced analytics, such as Microsoft's Power BI.
3. **Peak demand management.** Minimize usage in peak demand periods, to avoid triggering higher rates and potentially harsh penalty fees. This is done by classifying the real-time data into event frames, calculating projected usage, and using EIDI real-time notifications to generate warning alerts well before the peak thresholds are reached.
4. **Idle state management.** Reduce energy usage to absolute minimums when running assets in an idle state.
5. **Demand/response management.** If feasible, offer excess energy capacity back to the power grid when requested in exchange for revenue or incentives.

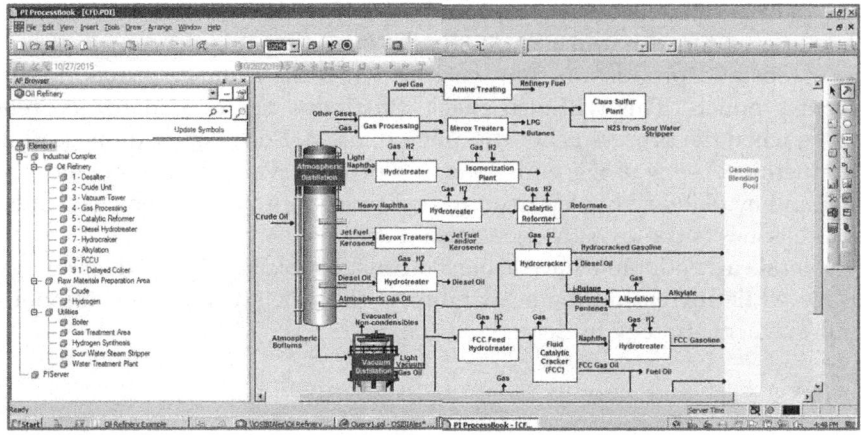

FIGURE 7.2
Integrated ProcIndustries refinery diagram with asset data model for continuous improvement.

In addition, management asked Peter to work with refinery teams to implement a plan to better ensure energy resiliency during storms or lightning strikes. Ron reviews their approach and further recommends that they capture energy data (electricity, fuel, and steam flows) for each unit, classifying the information in order to subdivide consumption by unit for subsequent analysis. He also wants to capture unit conditions and losses when the plant is not on target. Ron proposes integrating production information, energy usage, emissions data, and process availability in order to monitor the cost of electricity under specific conditions.

A detailed process flow diagram with the key energy centers of the industrial complex assists in identifying the energy areas to consider. Using a process block diagram adds the necessary context to the process, asset, and energy data.

Figure 7.2 shows a schematic of the South Texas refinery block flow diagram to gather all information from all process areas. All relevant inputs that can be measured should be included for treatment by the data infrastructure system.

Metering and Inputs

The team standardizes naming conventions for unit attributes when collecting and analyzing real-time data. It is also beneficial for data modeling, data exchange, and reporting purposes. They include weather data to look for seasonal patterns that can affect key performance indicators (KPIs) and detailed analysis. Some chemical processes are always affected by current atmospheric pressure, temperature, humidity, and wind speed. Energy consumption is sometimes dependent on raw material types and heating processes.

Data Capture and Reporting

Data must be mapped into the asset data templates from the meters to the programmable logic controllers (PLCs), supervisory control and data acquisition (SCADA) systems, and distributed control systems (DCSs).

Peter Argus and Monica Armstrong, the planning and economics coordinator, have created the required analytics to calculate total energy consumption, total water consumption, and other variables in their templates. Their data classification strategy simplifies data aggregation for identifying hidden losses in order to define process improvements that directly reduce refinery costs.

Figure 7.3 shows the refinery block flow and process flow diagrams including auxiliary equipment. The key energy inputs to the refinery are gas, fuel, air, and water required to produce electricity, steam, and compressed air to support the operations. The team also plans to look at gas emissions produced by energy transformation as well as water cooling and treatment.

Ron has been working with Peter and Monica to define the process unit object model, or asset template, for their refinery. Figure 7.4 shows the detailed data model to include all consumable variables in the process unit model. Monica provides the team with the basic model used in their planning and economics optimizer.

Data Analysis, Visualization, and Reporting

Data configuration, data analysis, and reporting efforts are reduced using reusable templates that feature real-time notifications when an exception

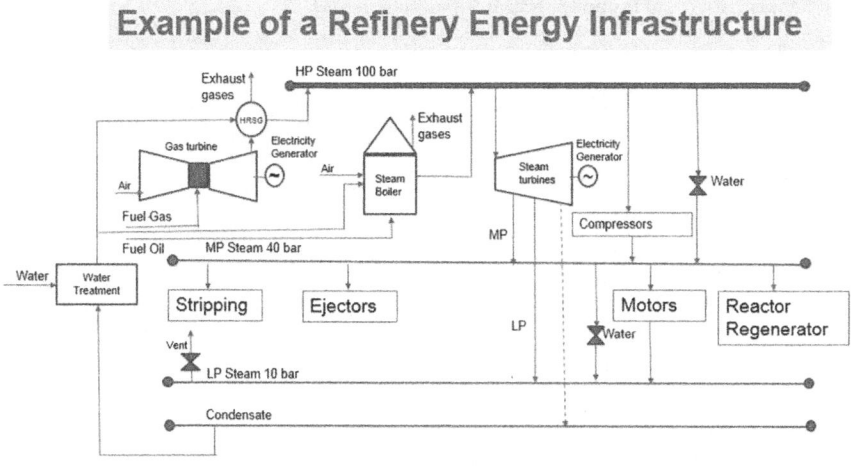

FIGURE 7.3
ProcIndustries simplified energy process diagram.

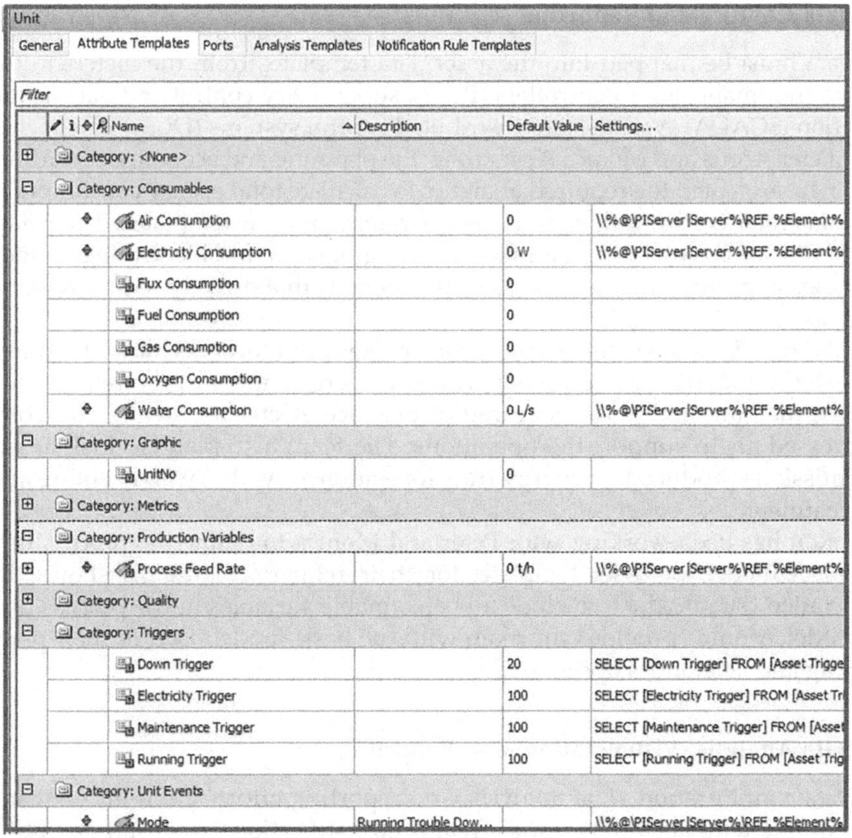

Unit					
General	Attribute Templates	Ports	Analysis Templates	Notification Rule Templates	

Filter			
✐ i ✥ ℛ Name	△ Description	Default Value	Settings...
⊞ 📄 Category: <None>			
⊟ 📄 Category: Consumables			
♦ 🔲 Air Consumption		0	\\%@\PIServer|Server%\REF.%Element%
♦ 🔲 Electricity Consumption		0 W	\\%@\PIServer|Server%\REF.%Element%
🔲 Flux Consumption		0	
🔲 Fuel Consumption		0	
🔲 Gas Consumption		0	
🔲 Oxygen Consumption		0	
♦ 🔲 Water Consumption		0 L/s	\\%@\PIServer|Server%\REF.%Element%
⊟ 📄 Category: Graphic			
🔲 UnitNo		0	
⊞ 📄 Category: Metrics			
⊟ 📄 Category: Production Variables			
⊞ ♦ 🔲 Process Feed Rate		0 t/h	\\%@\PIServer|Server%\REF.%Element%
⊞ 📄 Category: Quality			
⊟ 📄 Category: Triggers			
🔲 Down Trigger		20	SELECT [Down Trigger] FROM [Asset Trigge
🔲 Electricity Trigger		100	SELECT [Electricity Trigger] FROM [Asset Tr
🔲 Maintenance Trigger		100	SELECT [Maintenance Trigger] FROM [Asset
🔲 Running Trigger		100	SELECT [Running Trigger] FROM [Asset Trig
⊟ 📄 Category: Unit Events			
♦ 🔲 Mode	Running Trouble Dow...		\\%@\PIServer|Server%\REF.%Element%

FIGURE 7.4
Unit template for event frame analysis by operating modes.

is detected. Two other ideas added further granularity to the data: (1) capturing process events for data aggregation and (2) using target production values to identify minor operations delays that cause energy and consumable losses.

In addition, the template computation capabilities totalize data; convert values to their desired process engineering units; and normalize the efficiency calculations for conveyors, pumps, compressors, and rotating equipment. The hierarchy and template design simplify information access, providing desired context and detail. The original data at its original resolution is always available. Therefore, the design of the analysis layer determines how successful the applied model or abstraction is.

The EIDI data can be fed to non-operations displays, such as the Microsoft Power BI pivot table shown in Figure 7.5. This interactive consumption report is based on event frames that have captured unit-operating

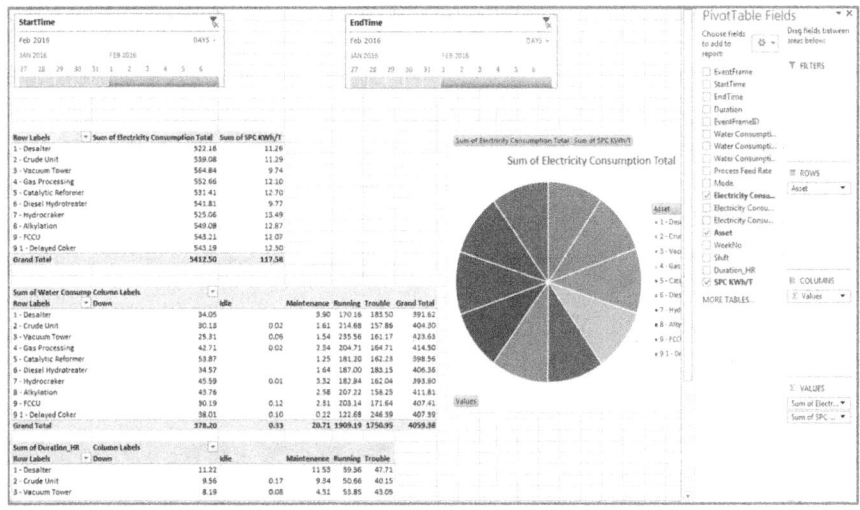

FIGURE 7.5
Excel report depicting real-time aggregation of EIDI production and consumable data by operating mode.

conditions. Total energy consumption is shown, filtered by operational events. The tables are generated by selecting the power table fields shown on the right-hand side of the figure. The table is composed by rows listing all assets, with the right-side column allowing the user to filter consumables and operating modes.

When building this Power BI report, Monica surmised that exporting and including all of the relevant data was key. She decided the user would filter information that the user is not interested in analyzing at a particular time. This approach is simpler and more prudent than not storing the available information, as the information may be needed for additional or future analysis. This also minimizes report maintenance, as we never know what information will be needed in the future for internal reporting, more focused analysis, or extracting information for external regulatory compliance or litigation. (Exporting EIDI data to advanced analytics such as Power BI is extensively covered in Chapter 10.)

A key is to have the event frame template for process units strategically designed for operational intelligence analysis. The event frame template can always be adapted to include additional information if required. The concept is to aggregate the data based on the current variance set by the production schedule. For example, additional consumables can include each type of steam, water, fuel gas, fuel oil, electricity, catalyst, hydrogen, acid, caustic, and other reagents as presented in Figure 7.6.

Figure 7.6 shows the classification of the electrical and water consumption for the five production states selected for analysis. As such, the EIDI has been

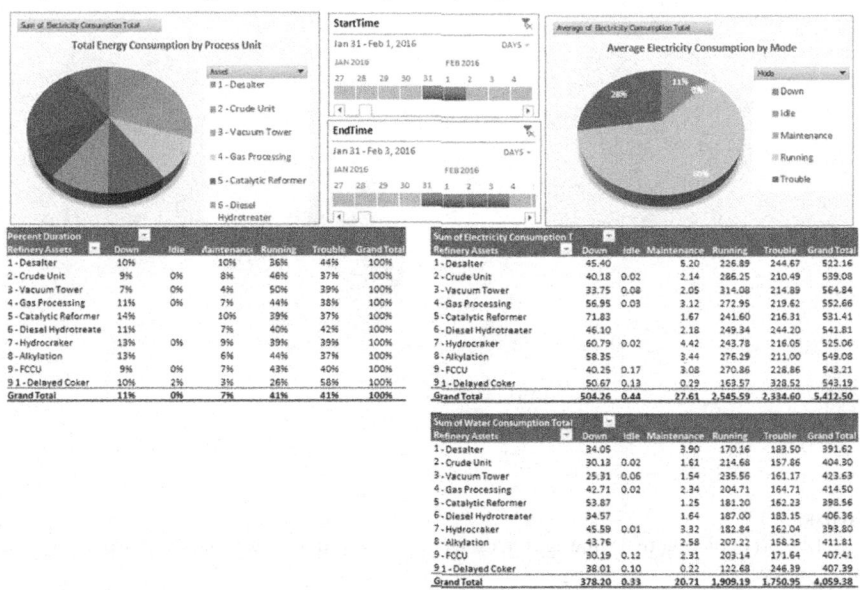

FIGURE 7.6
Microsoft Excel Power Pivot: Energy and water consumption.

assisting the engineers for all process units to classify the events for aggregation of the data for the five states. The results show that the minor losses caused by times when units are running in trouble mode is almost equal to the running times. These results indicate that there is room for improvement in analyzing the trouble events to find out why the operators have to slow down the units. (The data presented is not real; however, the process and generation of the interactive report is real using the suggested strategy and tools.)

The total energy consumption and the specific energy consumption are tracked in real time. By comparing these metrics with target values, the performance of the unit can be analyzed and opportunities for improvement researched if economically feasible. All of these calculations are performed in the EIDI analytics, and the event-framing strategy provides deeper insights to estimating losses caused by not meeting the targets because of operating troubles, idle time, equipment downtime, or scheduled maintenance.

Once the basic root causes of minor losses are identified and resolved, the team can consider optimizing operations.

Figure 7.7 displays 12 months of refinery energy consumption and feed rates, indicating that during July and August, the refinery was in turnaround mode.

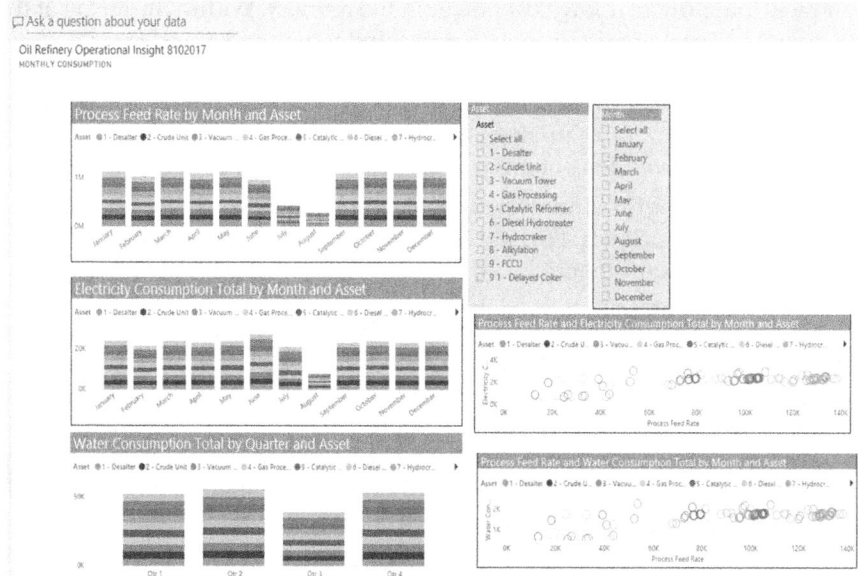

FIGURE 7.7
Power BI energy intensity monthly results for one year.

A Takeaway for the Team on Measuring Power Consumption

As the team continued to make progress on reducing and optimizing energy usage, they noticed that energy consumption by itself did not tell the entire story. They drilled down a bit deeper and discovered that power quality issues impacted several of the operating units and the data center. They began to closely monitor frequency, voltage, and current (amperage) to uncover momentary power interruptions. Assets like pumps, motors, and compressors have current curves based on discharge pressure, flow, or other performance variables. So, the team used voltage and amperage for resiliency analysis.

The team realized that consumption analysis could be more than simply assigning excess or unusual usage to a specific asset, batch, or shift. As a result, they discovered that energy metering is not simply consumption. Rather, they augmented their analysis to include consumption and power quality.

Smart Grid and Refinery Resiliency Improvements

The digital transformation team addressed management's desire to minimize unwanted outcomes from storms or weather events, such as lightning strikes. The South Texas refinery did not make its own electricity from an onsite power plant. During and after Texas had deregulated energy supply in the early 2000s, ProcIndustries' businesspeople had negotiated several power supply contracts that provided low-cost power to the refinery. Within the terms of the existing contract were provisions that ProcIndustries had asked for:

1. Some relief on power prices during peak summer days in return for the grid operator requiring cutbacks in refinery power usage; being flexible allowed for potential large cost avoidances.
2. The rights for ProcIndustries to monitor the substation that adjoined the refinery property, even though the local grid operator owned the substation equipment.

As the project progresses, the EIDI infers that more granular data can more accurately track power consumption usage for individual assets. As such, ProcIndustries deploys additional power meters to track real-time power consumption of all turbines, agitators, compressors, and pumps. They further calculate the total power consumption by plant area using the motors' nominal rating multiplied by the on-and-off digital switch that they had available through the refinery's DCSs and passing this data into the EIDI for online calculations. Each area of the plant is also equipped with an energy and water consumption usage meter. By tracking operating mode in the EIDI event-based data, they substantially improve the estimate of total energy consumption (energy, fuel, water, air, oxygen, hydrogen by area by unit) for each operating mode. In addition, energy usage decreases because of improved availability and performance of the equipment in the units. The engineers are able to reduce unit "trouble times" through improvement of process control strategies using real-time EIDI analytics and exporting the data to offline analytics software.

The next task was improving the resiliency of the refinery when unexpected power outages affected the refinery's ability to make product effectively because of an inability to have visibility into unit processes during those times.

Peter had attended OSIsoft user conferences for several years. One of his takeaways was how process and manufacturing companies had developed EIDI best practices for their own processes and applied them to other disciplines. For large facilities that made their own electricity, these mavericks used their EIDI systems to monitor their expected near-term power consumption and determine when to make their own power versus purchasing it from the grid. Even though the South Texas refinery didn't generate their own power, Peter decided to monitor the refinery's power and gas environment—storing custody transfer information—and local substation activity.

In addition to monitoring the health of the substation equipment using the same techniques described in Chapter 6, the team implemented the following:

1. They deployed a high-availability EIDI environment by adding a second EIDI server. In this configuration, both servers receive data, but one is the primary online server that facilitates EIDI users. If the primary server experiences issues that impact availability, all activity transfers

to the mirrored secondary EIDI, with users switched automatically in a matter of seconds. Once the primary server is restored, the information technology (IT) manager can revert back to the original environment. This approach also helps when applying cybersecurity updates and other related software patches to an EIDI system, as personnel can switch back and forth with disruption held to an absolute minimum.
2. The team worked with power subject matter experts and implemented a higher resiliency environment. By purchasing larger, higher capacity uninterruptable power supplies (UPSs), including larger capacity batteries, the refinery's control systems and EIDI systems (primary and backup servers) now have four hours of run time when the power unexpectedly goes out.
3. The team worked with the local power grid company to set up real-time monitoring of the substation located at the refinery gate. By deploying intelligent calculations based on input from the grid operator, they learned some of the metrics that forecasted a possible power interruption. This would give them early detection of an impending problem and provided the few minutes necessary to ensure that the UPS environment was fully running.

Using Process Flow Diagrams for Energy Management

Process flow diagrams with sensor data are the masterworks for translating measurements into valuable performance data. They can be used to infer variables based on heater, pump, and heat exchanger models and can be calculated in real time when the process unit-operating mode is set to running on target.

This means that the data is satisfactory for augmenting the knowledge of the sensors to estimate the metrics and performance calculations to increase the operational assessment of a process unit. The laboratory data and process data can also be reused using statistical methods for multivariate analysis and modeling. As such, the whole set of process data, laboratory data, and equipment data are used to identify if the unit can be moved to optimal conditions within the operating windows (Plourde et al. 2017; Steyn et al. 2018; Bascur and Soudek 2019).

As discussed earlier, energy represents 66% of the cost of running a modern refinery. Of this, the largest component is furnace fuel. Alex Moretti, refinery process engineer, prepared a list of all the pumps in the crude unit area with the online measured flow and head from the sensors. He also included the observed heat transfer coefficient (*observed U*) versus the design clean measurement (*design U*) to estimate the efficiency of the heat exchanger. He generated a spreadsheet showing which pumps required overhaul and which heat exchangers required cleaning. Table 7.1 shows an example of Alex's

TABLE 7.1

Performance Test Pump and Exchanger Efficiency

Pumps

Pump Number	Observed Flow	Observed Head	Design Head at Flow	Pump Overhaul?
P-101	32	38	60	Yes
P-102	105	202	200	No

Heat Exchangers (HE)

HE Number	Observed U	Design Clean U	Design Service U	Needs Cleaning?
E-101	77	104	63	No
E-102	11	98	61	Yes

EIDI spreadsheet report. This initiative augments the maintenance department's condition-based maintenance (CBM) program.

Alex was happy to discover that most of the thermocouples and flow meters were in place to do these efficiency calculations. Although the efficiency calculations could be done using the equations in the spreadsheet, now a single equipment template can be used for the design and reused for all the pumps and heat exchangers in the refinery. This simplified the analysis and provided results that are easily adaptable for many uses.

Figure 7.8 shows a smart template for heat exchanger fouling using EIDI run-time analytics. These online calculations monitor performance of

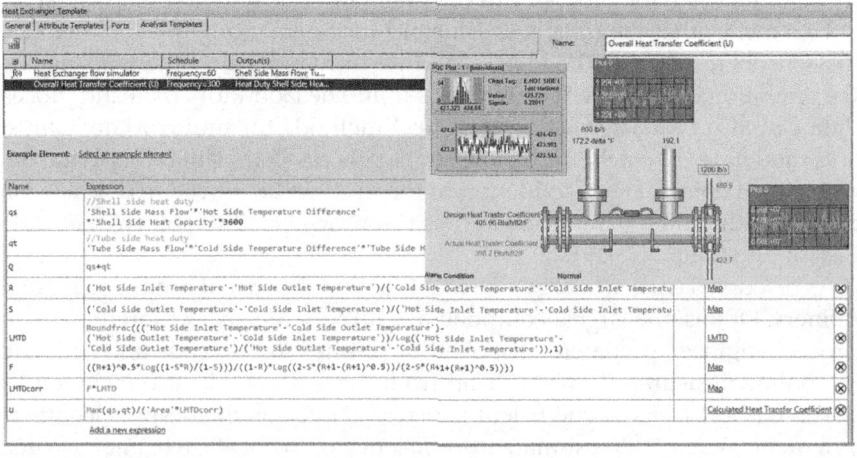

FIGURE 7.8
Smart asset template for heat exchanger fouling using EIDI run-time analytics.

hydraulic and energy calculations. It will enable continuous tracking of all the units in real time and will assist in satisfying environmental and safety constraints. As we have discussed, the key is actionable information and direct access to data in real time.

Using Advanced Analytic Tools to Gain Real-Time Insights

In Chapter 4, we presented how classifying the raw data using operational modes provides real-time insights with no additional software tools. We showed the value of creating a template that used basic production and consumable variables to characterize the operation. The EIDI is capable of aggregating the data for overall production effectiveness evaluation by unit and by shift.

The *running OK* mode data has shown to be extremely valuable in generating predictive analytics of quality variables based on the operational data captured from this data subset. Additional information generated for the running OK data reduces the complexity and dimensionality of the problem so it is easy to understand.

Principal component analysis (PCA) is a statistical tools method for dimensionality reduction. The multivariable operational data set, consisting of several laboratory variables and process variables, are treated with PCA. In process applications, the data are usually multivariate, collinear, noisy, and typically missing. This makes it difficult or impossible to use full-rank statistical methods for modeling and analysis, that is, multiple regression, discriminant analysis, analysis of variance, as well as neural networks. However, projection methods such as PCA and projection of latent squares (PLS) are based on realistic assumptions about the variables (collinear, noisy, etc.), making them suitable for modeling and analysis of complicated processes and other data. PCA and PLS analysis results can be displayed graphically, including multivariable control charts and multivariable statistical process control charts (MSPCs), for example, Shewhart, exponentially weighted moving average (EWMA), and cumulative sum (CUSUM) charts. These MSPC charts, combined with charts based on the model residuals, provide tools for early fault detection and the identification of drift, mean shifts, and so forth. Additional plots indicate the variables that are likely related to process problems upsets and other process events. Garrigues et al. (2000) shows additional ways to treat the data. And many others have provided valuable descriptions and examples in the chemical, iron, steel, and mineral processing industries such as MacGregor and Kourti (1998) and Dudzic (1998). These EIDI online diagnostic tools can also be implemented for both operations.

Mass Balances and Data Reconciliation

There are several requirements for developing a methodology to implement a data reconciliation system. First, the algorithm that is able to balance and reconcile the plant data must be robust and must perform correctly against any process topology or configuration. In the past, several mathematical and statistical tools have been developed to solve this reconciliation problem. However, a mathematical algorithm without an information infrastructure is of little value. Second, the right infrastructure to connect the object-oriented model to the real-time process data must be in place. A database that allows storage and manipulation of elements is necessary. The system has to adapt to changes in the process topology because, for example, a meter going out of service is enough to change the mass balance of a process network. Additionally, this object-oriented database should be able to communicate in real time with the EIDI system. Third, the EIDI system acts as a repository of process data, both raw and reconciled, that allows the distribution of data to all plant staff, from the operators to the plant manager.

The typical problems with process data in industrial plants are:

- An overwhelming amount of data;
- Low confidence in the available data;
- Lack of consistency (the data do not make sense);
- Data violating known constraints (mass and energy balances); and
- Poor data quality creating a decision-making fog at all levels of an organization, which may result in a financial penalty (fine).

In the past, these problems have been addressed using traditional methods to capture the required information. However, a key issue with these traditional methods is that human errors can occur during manual data entry, and operational events may be ignored.

Online process stream analyzers provide chemical assays for a given time period. The laboratory provides sample data for the streams. As such, these measurements provide redundancy, which require data reconciliation to find the measurement errors and close the mass balance. Having data redundancy enables us to infer other streams' properties, which are not measured. The unmeasured variables are estimated using a least squares reconciliation algorithm. The model (plant structure) can run hourly, by shift, or daily to produce reconciled reports. In addition, the reconciled data is sent to the EIDI system where it is available to operations support staff and knowledge workers. Figure 7.9 describes the daily procedure to reconcile process data. The unbalanced data is collected from the EIDI. Once the data is in the system, a group of analysis rules is executed to detect gross errors that can negatively affect results. These

Energy Management and Operational Improvements

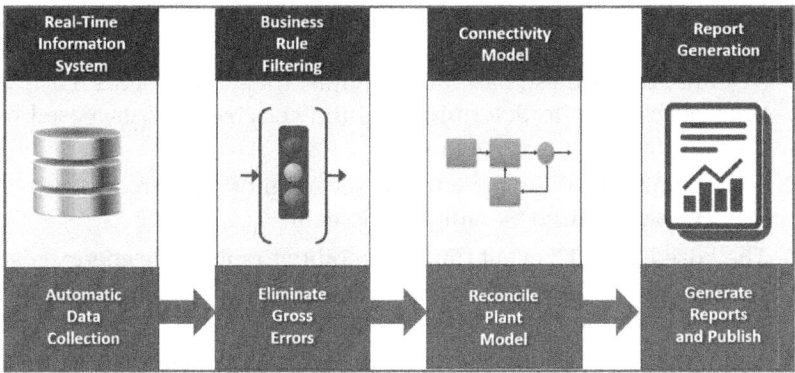

FIGURE 7.9
Daily procedure for data reconciliation. (From Bascur and Linares 2005.)

gross errors are eliminated before the final reconciliation run, which provide a unified balance for the whole complex.

An oil refinery mass balance model includes the crude receipts, storage tank inventories, and the shipping and loading rack receipts. Once the mass flow, stream, and inventory compositions are reconciled by the system, many calculations and yield reports can be performed. It is important to realize that in the process of solving the network problem, the measurements are validated and gross errors are detected prior to providing a solution.

Once the data are reconciled, they can be sent to the enterprise resource planning (ERP) and other business information systems, where they can be distributed to users. The infrastructure of the data reconciliation system has to adapt to any changes in process flow or changes to the measuring system because both the process topology and the data are not static.

The reconciled data can then be used to improve the process to continuously improve yield performance by looking at the overall refinery performance. The data can be reused to improve process planning and to determine the optimal setpoints for steady-state optimization of the plant (Narasinham and Jordache 2000; Bascur and Linares 2005). Hodouin (2011) also described additional uses for this type of methodology.

What You Should Take Away

EIDI data can be used and reused for many different use cases, including energy management and improving facility resiliency. In this chapter, we saw Ron Erickson and the refinery transformation team take the following steps to investigate how best to capture and analyze information that assists in their quest for optimum energy consumption:

1. They monitored total refinery energy usage (power, water, gas, steam) for one month in order to establish an energy consumption baseline. Using these baseline amounts, they developed EIDI runtime calculations to determine real-time electricity costs, based on the current power price of power.
2. They continued their proactive asset monitoring to reduce excess energy usage caused by faulty equipment.
3. They used the EIDI event frame capability to subdivide energy costs by unit, product, sub-process, or other event to determine where and when the largest costs are incurred.
4. Using these calculated costs, real-time energy cost monitoring generates alerts to the operations team when energy consumption starts to exceed expected usage or when energy usage approaches a limit whereby the company will pay a punitive fine for exceeding consumption during peak seasonal periods.
5. They further analyzed energy usage to determine which operating states caused the biggest usage, relative to expected consumption.
6. They strengthened their EIDI and their facility's energy resiliency so they are not totally dependent on the local power grid. By implementing a secondary server for 99.9% high availability of EIDI data and strengthening their UPS environment for their refinery control network, they are better equipped when storms or other power problems are looming. Specifically, they will not suffer data visibility outages in the refinery as they had in the past.
7. They improved resiliency by developing EIDI calculations that notify refinery personnel when outside power supply from the power grid might be compromised so they can augment power by using UPS equipment.

We also described how the use of process flow diagrams can assist process engineers by developing EIDI inline calculations that determine pump and heat exchanges performance, which aids the maintenance team in making their CBM initiatives more effective. Finally, we described how tools, such as mass balance calculations and data reconciliation can improve refinery operations when used with an EIDI system.

References

Bascur, O.A., and Linares, R. 2005. "Grade recovery optimization using data unification and real time gross error detection." *Mineral Engineering* 19(6–8):696–702.

Bascur, O.A., and Soudek, A. 2019. "Grinding and flotation optimization using operational intelligence." *Mining, Metallurgy & Exploration* 36(1):139–149.

Dudzic, M. 1998. "The use of multivariable statistical technologies." Presented at the AISE Specialty Conferences, Boston: MIT.
Garrigues, L., Kettaneh, N., Wold, S., and Bascur, O.A. 2000. "Multivariate process analysis and optimization in mineral processing." In *Control 2000*, ed. J.A. Herbst. Littleton, CO: Society for Mining, Metallurgy & Exploration. pp. 41–50.
ISO 50001. Energy Management. www.iso.org/iso-50001-energy-management.html. Accessed August 2019.
MacGregor, J.F., and Kourti, T. 1998. "Multivariable statistical treatment of historical data for productivity and quality improvements." Presented at FOCAPO 98, Snowbird, Salt Lake City.
Natural Resources Canada. 2019. Energy management for industry. www.nrcan.gc.ca/energy-efficiency/energy-efficiency-industry/energy-management-industry/20397. Accessed August 2019.
Plourde, M., Bascur, O.A., Paquet, S., and Gervais, D. 2017. "Digital innovation in modern engineering and operational excellence." Presented at the 2017 SME Annual Conference and Expo, Denver, February 19–22.
Steyn, J., Bascur, O.A., and Gorain, B. 2018. "Metallurgy analytics: Transforming plant data into actionable insights." *Mining Engineering* 70(9):18–29.

Additional Reading

Bascur, O.A., and Aroqui, R. 2014. "Drastic reductions in energy and water consumption in large iron and steel metallurgical complexes." In *AISTech Proceedings*. Pittsburgh, PA: AIST.
Bascur, O.A., and Halhead, M. 2013. "Energy effectiveness and sustainability management at Anglo American Platinum." In *Proceedings Copper 2013 International Copper Conference*, vol. 1, eds. C. Moscoso, J. Rosales, and A. Vio. Santiago: Chilean Institute of Mining Engineers (IIMCH). pp. 415–423.
Bascur, O.A., Halhead, M., Garrigues, L., and Jarvis, M. 2016. "Mineral processing plant asset and energy optimization: The calming cloud over operations." In the *International Mineral Processing Congress (IMPC XXVIII) Proceedings*. Quebec City: IMPC.
Beal, J. 2016a. "Loop tuning basics: Integrating processes." *Intech Magazine* (March/April): 48–51.
Blanchard, B.S., and Verma, D.C. 1995. *Maintainability: A Key to Effective Serviceability and Maintenance Management*, 2nd rev. ed. New York: Wiley-Interscience.
Cope, K., and Chugh, S. 2019. "Predictive analytics for reliable high power performance, Ecolab Nalco Water." Presented at PI World, San Francisco, CA.
Efficiency New Brunswick. 2010. *Energy Management Information Systems: Planning Manual and Tool*. Ottawa: Natural Resources Canada.
Ferrari, A., and Russo, M. 2016. *Introducing Microsoft Power BI*. Redmond, WA: Microsoft Press.
Harjunkoski, I., Scholtz, E., and Feng, X. 2014. "Smart grid: Industry meets the smart grid." *Chemical Engineering Process* (August): 45–49.
Johnson, D. 2015. "Pervasive sensing saves money and prevents downtime." *Intech Magazine* (July/August): 66.

Krenek, D. 2011. "Improving reliability with Caterpillar condition monitoring services." Presented at the OSIsoft Users Conference, San Francisco.

Morrow, M., and Zhang, B. 2019. "It's the data, stupid: How PI AF analytics rescued our real-time multivariate process monitoring." Presented at OSIsoft PI World, San Francisco.

McKinsey Global Institute. 2015. *The Internet of Things: Mapping the Value beyond the Hype*. San Francisco: McKinsey & Company. www.mckinsey.com.

Rossiter, A.P., and Davis, J.L. Jr. 2014. "Identify process improvements for energy efficiency." *Chemical Engineering Progress* (September): 53–58.

Ruel, M. 2010a. "Closed loop tuning vs. open loop tuning: Tuning all your loops while the process is running is now possible." bba.ca/publication.

Ruel, M. 2010b. "Control system performance assessment—Best practices." In *Proceedings of the Second International Congress of Automation in the Mining Industry (Automining 2010)*. Gecamin, Santiago. p. 54.

Ruel, M. 2014a. "Advanced control decision tree." In *Proceedings 46th Canadian Mineral Processors*. Westmount, QC: Canadian Institute of Mining, Metallurgy and Petroleum.

Ruel, M. 2014b. "Successful methodology to select advanced control approach." Presented at the Process Control and Safety Symposium, International Society of Automation, Houston, October 6–9.

Scheihing, P. 2014. "Save energy through the superior energy performance program." *Chemical Engineering Progress* (September): 48–51.

Smith, C.L. 2016a. "PID explained for process engineers: Part 1: The basic control equation." *CEP Magazine* (January): 37–55.

Tomlingson, P.D. 2009. "Reliability-centered maintenance improve equipment health." *Mining Engineering*, pp. 41–49.

Van Scholkwyk, T. 2012. "PI applications in Anglo American Platinum, control loop monitoring and reporting." Presented at the OSIsoft Regional Seminar, Johannesburg.

Weber, C., and Yeung, T. 2010. "Energy management raising profitability, reducing CO_2 emissions." *Petroleum Refining Quarterly* 4:87.

8

Successful Examples of Enterprise-Wide Digital Transformation

Steve Jobs gave a small private presentation about the iTunes Music Store to some independent record label people. My favorite line of the day was when people kept raising their hand saying, "Does it do [x]?," "Do you plan to add [y]?" Finally Jobs said, "Wait, wait—put your hands down. Listen: I know you have a thousand ideas for all the cool features iTunes could have. So do we. But we don't want a thousand features. That would be ugly. Innovation is not about saying yes to everything. It's about saying NO to all but the most crucial features."

Derek Sivers

Chapter Overview

We take a short break from the ProcIndustries South Texas refinery and their continued use of the enterprise information data infrastructure (EIDI). In this chapter, we describe how actual process manufacturing companies have transformed their companies through innovative uses of their own EIDIs (e.g., OSIsoft PI System). Although these companies are in different industry sectors, their use cases, methodology, and results are strikingly similar.

The following company use cases are presented:

- **An international integrated oil and gas company that we call OilCo.** The example presented focuses on their use of EIDI (PI System) data in downstream refining.
- **A North American midstream petroleum services company that we call MidPetCo.** The use case presents improvement of their pipeline operations through data analysis.

- **An international gold-mining company that we call GoldMineCo.** The use case describes how intelligent use of operations data resulted in a lower cost of producing gold.
- **A multinational building and construction materials company that we call MaterialsCo.** The use case presents how they are monitoring and automating their fleet of plants.
- **An international specialty chemicals company that we call ChemCo.** We present how ChemCo uses data throughout the company to solve difficult, hard-to-diagnose production issues.

You will see how these companies went about their digital transformations and how they prioritized their use cases, which directly correspond to the material we presented in earlier chapters of this book. We found some common threads in how customers leveraged their own EIDI:

1. The EIDI became their consolidated time-series system of record.
2. Existing data silos were consolidated, including a multitude of personal Excel spreadsheets.
3. They added context to the EIDI data in real time, so that it became more easily usable by people or analytics software.
4. The companies enabled self-service access to the EIDI data.
5. They collaborated to solve difficult problems and to innovate with new use cases, changing the culture.

OilCo—Use of EIDI Data in Downstream Refining

OilCo is a multinational, integrated oil and gas company, consisting of numerous refineries and petrochemical plants across several countries. OilCo also owns 2,000 petrol filling stations in 13 countries. Several of its refineries are among the most profitable on their continent.

OilCo has transformed the way they refine energy and produce petrochemical products by better leveraging operational data collected by their distributed control systems (DCSs) and other refinery and plant data collection systems. OilCo's EIDI has become their universal operational technology (OT) language and generates more than 80 billion data points annually.

Starting in 2012, when faced with significant market and operational challenges, OilCo embarked on their digital transformation—New Downstream Program (NDP)—using the following fundamental strategies:

1. OilCo viewed their operational data as a strategic asset requiring a fit for purpose infrastructure—the EIDI.
2. They turned the operational data into operational intelligence and put the data in the hands of their users via self-serve access, thus distributing in context to the largest audience possible, so that people in various roles could use their talents to make product more efficiently and at a lower cost through continuous improvement.
3. They began using a *layers of analytics* strategy that first defined what "analytics" meant to OilCo. To do this, OilCo developed an analytics framework, defining types of analytics and determining the best "fit for purpose" technology to implement these analytics. The layers of analytics were leveraged to underpin an analytic strategy that determined what type of analytics should be used to solve a particular business need. OilCo then defined the following analytics framework:
    a. Descriptive
    b. Diagnostic
    c. Level 1 Predictive—predictions based on empirical correlations, first-principal equations, or models
    d. Level 2 Predictive—predictions based on pattern recognition, machine learning, or other advanced analytics method

    OilCo used the EIDI's capabilities to perform real-time, streaming analytics via the digital replicas in a template format, which was developed and maintained by their subject matter experts (SMEs). In this approach, the EIDI served as the foundation for all analytics. People have immediate access to key production indicators (KPIs) and asset health statistics and are notified at the earliest possible moment when conditions deviate from normal or optimum ranges. Contextualizing the data in an online environment reduced the need to prepare and confirm data accuracy when exporting the data to offline analytics and models.
4. OilCo also defined levels of information technology (IT)/advanced analytics into technical (unit level), strategic (refinery level), and community-focused levels to enable digital value chain integration with key suppliers such as their catalyst vendors.

OilCo's EIDI Deployment

In 1998, OilCo began their digital journey by initiating the use of an OSIsoft PI System as their EIDI, as a *data historian*, and have constantly expanded their EIDI scope to a combined total of more than 400,000 data points. In 2012, OilCo upgraded to a modern PI System and focused on fully leveraging PI Asset Framework (PI AF). PI AF enables a hierarchical metadata structure that defines all of OilCo's physical assets and their corresponding components, including sensor data values and inferred run-time calculated data values. OilCo has used this heavily and today has more than 300 AF smart asset object templates, 21,000 object template elements, and over 60,000 event frames that define and signal the occurrence of key process segments or deviations from normal operations. In short, OilCo is converting vast amounts of operational data into exception-based operational intelligence.

The EIDI has served as the operations system of record. In 2010, OilCo's Information Integration and Automation team, working with their SMEs, configured a *digital twin* of different processes and equipment sets in a facility. All of the relevant data, calculations, analytics, and notification alerts are combined into a comprehensive digital replica of the physical plant in asset class templates, so it can be more easily visualized and analyzed. The templates enable rapid deployment and manageability as any changes to the templates by the SMEs are propagated across the enterprise where these templates are used.

Using this digital twin as the centerpiece, OilCo provided self-serve visualization tools so their employees could view and analyze the operations and production information through configurable KPI dashboards, asset relative graphics for plant and unit overviews, and historical and event data displays for root cause analysis. By quickly configuring and customizing contextualized displays to fit people's specific roles, time to insight was reduced. For personnel who needed data in Microsoft Excel, the EIDI provided a standard pull-down menu within Excel to import EIDI real-time sensor data, calculated and totalized data, historical data, or statistical data over a given time range. This was useful for predefined internal reports, compliance reports, or ad hoc analysis, where the EIDI data could be exposed to math and other Excel add-ins.

OilCo also built a palate of self-serve analytics and business intelligence where operators, engineers, and businesspeople configured their own smart asset objects, combined them like Lego blocks, and created a plant or process replica in which they could experiment in a development environment. Once completed, they implemented these enhancements across the OilCo enterprise with appropriate governance. Typical refinery use cases are shown in Figure 8.1.

OilCo reduced its IT costs and reliance on corporate IT personnel and outside vendors because employees were able to quickly build their own functionality on top of their EIDI infrastructure and then replicate it across assets throughout the enterprise. In doing so, OilCo simplified and standardized their application and solutions portfolio. Different data streams are able to be analyzed in tandem so that OilCo can determine the full impact (financial, maintenance, energy consumption) on changes to output.

Successful Examples of Enterprise-Wide Digital Transformation

Downstream OT Applications Using EIDI Data			
Safety and Asset Integrity	Process & Yield Improvement	Process/Production and Scheduling Optimization	Energy Management and Optimization
• Maintain Operating Envelopes • Advanced Alarm Management • Interlock Governance and DCS Tracking • Integrity Operating Windows (IOWs) • Proactive Asset Management Reliability for All Critical Rotating Equipment and Hydrogen Pressure Swing Absorbers	• Crude Blending Control • Yield Optimization & Reporting • Product Quality Improvement • Analyzer Reliability	• Planned vs. Actual Analytics • Natural Gas & Fuel Demand Gas Forecasting • Analyzing Normal Mode of Control Loops • Advanced Process Control Monitoring • Use of Pimsoft SigmaFine for Yield Accounting and Material Movements	• Real-time Energy Mgmt. • Energy KPIs (6 tiers) • Column Energy Efficiency Dashboards • Hydrogen, Utilities, and Energy Balances • Peak Electrical Forecasting • Flaring

FIGURE 8.1
Refining and petrochemical applications using EIDI data. (From Harclerode, C., "Data operations transforms fuels value," *PTQ Magazine*, Q1, 2017.)

Advanced Analytics and Machine Learning Leads to Significant Return on Investment

Once OilCo developed its smart OT infrastructure across their value chain, with the EIDI as the centerpiece, OilCo's focus turned to machine learning and big data analytics. They became one of the first large refiners to adopt Microsoft's Azure Machine Learning in a production environment. Microsoft Azure works in conjunction with their EIDI (PI System) operational data, which is uploaded to the cloud using OSIsoft's standard software integrator for offline business analytics. This integrator made it much quicker and easier for OilCo to extract, cleanse, and contextualize their EIDI data so that it could be quickly ingested as published data sets by Azure's Power BI and Machine Learning cloud-based analysis tools. A data flow diagram for a typical integrated oil and gas company is shown in Figure 8.2, with data sources on the bottom and business use cases on the top of the diagram.

OilCo developed Azure Machine Learning to predict the impact of sulfur levels in various feedstocks in their desulfurization units. OilCo had been using offline models to analyze the sulfur content. Not only did using offline models increase time, it also increased the potential for error. OilCo estimated it was losing more than US$500,000 annually because of its inability to adjust unit parameters that optimize sulfur content in products. OilCo eliminated those losses, thanks to better forecasting, and continues to hone its use of machine learning. The company has rolled out this technology across their enterprise for other machine learning use cases. As with its other

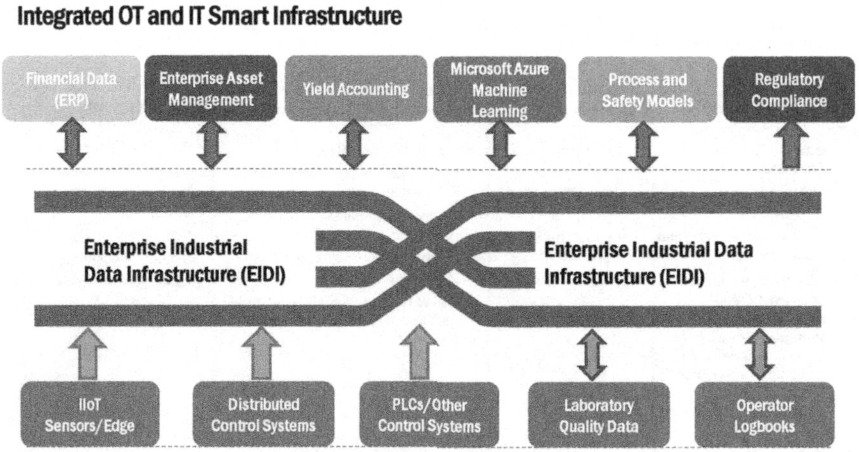

Integrated OT and IT Smart Infrastructure

```
[Financial Data (ERP)] [Enterprise Asset Management] [Yield Accounting] [Microsoft Azure Machine Learning] [Process and Safety Models] [Regulatory Compliance]
                                    ↕            ↕            ↕            ↕            ↕            ↕
       Enterprise Industrial                                          Enterprise Industrial Data
       Data Infrastructure (EIDI)                                     Infrastructure (EIDI)
                                    ↕            ↕            ↕            ↕            ↕
[IIoT Sensors/Edge] [Distributed Control Systems] [PLCs/Other Control Systems] [Laboratory Quality Data] [Operator Logbooks]
```

FIGURE 8.2
Data sources and integrated analytics with an EIDI as the pivotal component.

improvements, OilCo was able to leverage its previous technology investments and reuse EIDI data, so that this new application was layered on top of what had already been implemented.

Following these successes, OilCo turned to improving the performance of its delayed coking units (DCUs). By using opportunity crudes, the company estimated that it could realize a gain of US$6 million for each 1% gain in DCU yield. Gains in DCU yields with variable feeds from opportunity crudes, however, also increased the risk of steam explosions during the hydrocutting step.

Azure analytics, combined with continuous EIDI data feeds enabled OilCo to find the sweet spot. DCU yields were increased by 2%, yielding an estimated annual gain of more than US$500,000 per year for each unit. Concurrently, steam explosions were reduced by 75%. Machine learning enabled the achievement of two seemingly contradictory goals at the same time. OilCo positioned machine learning for its four DCU units across the enterprise to take full advantage of opportunity crudes.

On-Premise versus Cloud Analytics? No: On-Premise Plus Cloud

When using cloud-based machine learning analytics, data gets transferred and stored in the cloud. However, cloud systems will not replace on-premises storage systems. Real-time control and insight are required for operational efficiency and safety of the facility. With inherent latency with transferring data to and from the cloud, along with network disruptions, it is not advised to link real-time control of the plant/refinery with real-time cloud connections. It is much more prudent to have them complement each other by performing operations and production functions during peak production times and using the cloud or a digital twin for analytics to optimize operations. Using standard software tools to extract, cleanse, contextualize, and publish the OT data in a format easily digestible by the analytic tools significantly decreases time to insight when using cloud-based analytics. Some companies have reported 70% reductions in the time needed to prepare, gather, cleanse, and augment data needed for analytics.

By having the EIDI transform the collected sensor data into contextualized operations, production, equipment, and energy information, the path to having business systems and IT analytics leverage the refinery/plant information is substantially reduced by already understanding the information that will be sent to systems and analytics. Examples are

- KPIs that don't have to be interpreted and checked for outlier data, because the online KPI calculations have already done that;
- Separating data and outcomes from different types of production events, such as refining a specific grade of crude;
- Not having to manually totalize items, such as consumables, after the fact; and
- Knowing which data streams need to be analyzed and correlated, as opposed to a data scientist having to make assumptions about raw data coming from a data lake.

OilCo's Return on Investment

One of the compelling aspects of OilCo's digital transformation is that use cases can be added as the business needs them, without disruption or change to the existing infrastructure. The hallmark of a successful digital transformation is being able to continually turn operational data into operational intelligence, leveraging visualization tools, and analytics for applications that cannot be imagined today.

Using that strategy, OilCo estimates that its digital transformation program accounted for a savings of US$500 million in EBITDA (earnings measured before deductions for taxes, interest expenses, depreciation, and amortization) over the first two-year period from 2013 to 2014. During the next two years as part of their NDP (2015–2017), OilCo realized an additional US$500 million in EBITDA, bringing the four-year total to US$1 billion. While that rate of return may not continue, the company expects significant savings in the future from additional improvements through better understanding of its data. One of OilCo's current initiatives is increased product yield through better crude processing. Other initiatives that they are pursuing are flare gas recovery, hydrocarbon loss management, and of course, continuous improvement in all areas of operations.

Lessons Learned

Why has OilCo been successful where others have not seen quite the returns from similar efforts? OilCo believes the reasons are as follows:

1. The main goal is to deliver business value, not lead with technology. This strategy reduces costs, enhances safety, improves yields, and provides for more effective stewardship of their assets. Using technology and the latest analytics without business goals and returns is of little interest and unproductive.
2. They began their journey by creating a reliable, scalable, digital infrastructure that collected many types of real-time data, eliminating existing data silos and turning vast amounts of operational data into operational intelligence.
3. They enabled their SMEs, who have the deep understanding of assets and operations, to configure, in an evolutionary way, digital replicas of physical assets in templates that formed the foundation to layers of analytics strategy.
4. They understood their EIDI would be the bridge between aggregated, contextualized OT data and the IT analytics and business world. They automated this bridge through intelligent automation and integration of accurate OT data with business-helpful analytics.
5. They properly determined that contextualization of their real-time data be performed at the EIDI level rather than when the data reached offline analytics. As a result, they deployed OT calculations and totalizations at the OT level.
6. They see the increasing value of industrial Internet of things (IIoT) sensors to augment knowledge of their operations and feed their analytics more robust and highly granular data sets.

Use of EIDI Data by MidPetCo for Pipeline Operations

MidPetCo is one of the largest North American midstream oil and gas services companies, providing natural gas gathering, processing, and transportation services, operating in numerous U.S. states since before the Second World War. MidPetCo is also one of the largest producers of natural gas liquids (NGLs). They manage over 50 gas and fractionation plants, over 1,500 compressors, and over 50,000 miles of pipeline. These gas plants process billions of cubic feet of natural gas per day delivered by more than 50,000 miles of pipeline. This operation requires effective monitoring to ensure that natural gas, NGLs, and associated networks meet quality, reliability, safety, and environmental standards.

Overview and Challenges

In 2015, MidPetCo's management realized they had not significantly changed their operations in several decades. After a market downturn in which they were forced to do more with less resources, they began with executive leadership to ideate and gather information about leveraging digital technologies as part of an overall operational excellence focused business transformation. A small team of three people visited many different companies in sectors outside oil and gas to listen and educate themselves about best practices and lessons learned. One of the keys was the appointment of a chief transformation officer (CTO) with extensive and successful executive operations experience. The CTO then formed an organization consisting of both traditional IT, OT, and other nontraditional skill sets then felt necessary to lead and support transformative initiatives.

After extensive employee engagement, including ideas for improvement and visions of a transformed company, their reimagined strategy and supporting initiatives were formalized and kicked off. MidPetCo 2.0 focused on the operational excellence dimensions of people/culture, work processes, and technology to develop and implement transformative business results in asset and organizational efficiency, asset reliability, and risk management. Much of MidPetCo 2.0 is accomplished through adoption of digital technology and turning large amounts of tag-based operational data into self-service, contextualized operational intelligence.

As MidPetCo 2.0 began, existing systems were siloed and focused on the operations group, with a lesser emphasis on reporting and analytics. The company realized they required a single, contextualized view of all operations, with real-time visibility and alerting when items needed quick attention. As a result, they kicked off a five-year digitization plan in which they defined very specific and measurable goals of lowering their cost structure and enhancing collaboration in order to operate more efficiently and deliver differentiated customer service and value.

The first key decision was to create an Integrated Collaboration Center (ICC) for collecting and processing data across its operations. The second component was to create the MidPetCo Energy Lab, which sought to enable employees to successfully use the new digital technology and to evaluate and improve MidPetCo's current work processes through employee feedback.

MidPetCo 2.0 Initiative Leverages Operational Data for Profitability

In late 2016, MidPetCo entered into an agreement with OSIsoft to deploy PI System software as a real-time critical operations data integration, applications, and analytics infrastructure throughout their enterprise. They had been generating an enormous amount of digital operations data, but it was underutilized. They were confident after their research and engagement with OSIsoft that the PI Systems would assist in converting their raw data to actionable intelligence. Using an OSIsoft integration partner, MidPetCo's initial use cases were

- Implementing a consolidated real-time system of record,
- Creating broad situational awareness of fleet operations for real-time decision-making,
- Aggregating operator-round information with control system sensor data, and
- Maximizing equipment performance so that unscheduled downtimes are reduced.

An important part of MidPetCo 2.0 was the high-level linkage of digital strategy with business strategy and the determination of baseline KPIs with institutionalization of measuring for continuous improvement and reporting to their board of directors. These measurable business strategy KPIs, which tied back to EBITDA improvement targets, were to

- Increase ethane recovery,
- Reduce the operations and maintenance expenditures as a percentage of installation base,
- Reduce miles driven by their outside operational and maintenance personnel,
- Improve return on capital employed (ROCE), and
- Reduce environmental compliance and reporting costs.

To enhance their implementation, MidPetCo made use of the PI System's PI AF capability to create a representation of their fleet-wide assets. One of the keys to success was a strategy and approach to train and enable their SMEs to configure in an agile, evolutionary way, digital replicas (twins) of physical assets templates for real-time analytics, alerts, and notifications. These smart asset class templates were used to build out a digital replica (twin) of physical systems such as gas plants and compressor stations. In short order, MidPetCo had more than 500,000 data streams monitored by their PI Systems, along with more than 8,000 AF elements using over 400 templates (29 times scale), and more than 300 role-based displays across their enterprise, most of which were configured by the end users leveraging the PI AF asset hierarchy in a drag-and-drop basis. They also utilized PI event frames to segment their gas processing into significant, trackable events, comparing performance among various events. The use of templates allow for massive scalability and sustainability. This is illustrated in Figure 8.3.

Successful Examples of Enterprise-Wide Digital Transformation

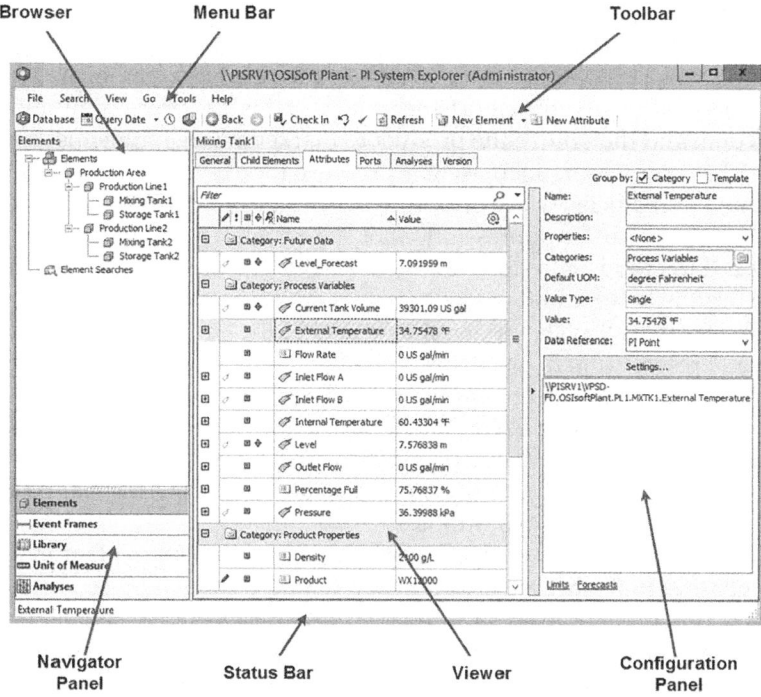

FIGURE 8.3
The power of PI AF templates in scalability and sustainability.

Real-Time Visibility Enhances Decision-Making and Reduces Inefficiencies

As a result, MidPetCo now possessed a real-time hierarchical view of their operations, coupled with contextualized data that crews were able to visualize and to make real-time, data-driven decisions. Concurrently, the ICC monitored overall operational performance and provided critical feedback to operations personnel.

MidPetCo wanted to use their data for gas plant optimization, which relies on several compositions. They integrated actual operating data with optimization models that simulated optimal performance, tracking and monitoring deviations between actual and optimal operating data, providing operations a view when the two diverged. By linking operational and simulated data to financial information, the operators were able to visualize a monetary value to any changes they made.

MidPetCo also defined and deployed real-time notifications to alert them when an asset diverged more than 10% from its optimal performance. This enabled the ICC and operations to perform root cause analysis and understand why it happened. As a result, they were able to submit reason codes for the equipment degradation or downtime. This enabled operations people to collaborate and improve the bottom line.

More than 2,000 MidPetCo compressors and pumps gather natural gas from the well heads, transport the gas to plants for processing and NGL extraction, and then move the product to market in large transmission pipelines. The compressors increase gas pressure for expedient travel, and their reliability and efficiency are critical to the business running smoothly.

To ensure that the compressor stations are running optimally, field crews make rounds to visually inspect them. Inspecting these stations takes time when driving to them and not knowing where problems might be located. By viewing their EIDI system data, they travel to the correct sites and can pinpoint and resolve urgent problems right away, reducing potential downtimes.

Today, MidPetCo has leveraged their PI System in real time and historically archived contextualized data to transition to condition-based and predictive maintenance practices. In addition, MidPetCo has developed a smart gas plant solution, in which they are now able to perform financial-based gas plant optimization every 15 minutes.

MidPetCo was able to create a dashboard that allows gas plant managers to compare actual margin performance versus potential margin in real time. MidPetCo maintains that, on average, if each of its 50–60 gas plants could produce an additional US$2,000–$5,000 per day through data-driven optimization, the potential gain is at least US$50 million annually.

MidPetCo also can shape its product mix, such as increasing ethane recovery due to market conditions, by examining various scenarios in real time. Having access to timely data on the price of ethane, contract terms, and plant operations is critical in a dynamic market such as gas processing. Timely decisions cannot typically be done without accurate, consolidated, information. The days of personal spreadsheets are winding down. In addition, companies must be able to prove that they are environmentally responsible. This requires accurate and readily accessible information for internal and external compliance and audits.

MidPetCo's Return on Investment and Future Plans

The company was able to meet the main objectives of MidPetCo 2.0:

1. MidPetCo realized a sizable return in one year (2017) on its investment in people, hardware, the PI System software, and deployment costs. The company achieved sizable returns in 2018.
2. They established a more efficient operation, featuring real-time situational awareness, and better, quicker decision-making. This included a successful deployment and continued innovation from its ICC.
3. The company successfully navigated their digital transformation journey through effective use of real-time infrastructure software and innovative use of data to achieve strategic corporate objectives,

such as asset and process optimization, integrating financial data into operations, and responsible environmental stewardship.
4. Employee productivity was enhanced through effective KPI dashboards, operations graphics with drill-down capability, and displays containing financial information so that operations workers see the economic impact of their decisions.
5. MidPetCo reduced avoidable maintenance costs and eliminated capital spending where data analysis sufficed.
6. The company's culture was transformed to become data driven. The *Imagine MidPetCo* program encouraged employee feedback on how workflow and collaboration could be improved. With that, the visualization and data entry technology MidPetCo implemented was supported by their workers. This includes a Facebook application for collaboration.

With a more efficient operation in place, company leadership intends to build on what they have achieved by

1. Developing additional real-time-based analytics for the gas plants and pipelines, including higher-level predictive analytics, such as advanced pattern recognition;
2. Linking operational data with geographical data and geospatial maps, so they can optimize gas routings across 50,000+ miles of gathering pipelines;
3. Exploring IIoT-enabled advanced machinery analysis to leverage more detailed collection of data from compressor units, including developing smart booster stations, enabled by real-time data displayed from virtual reality (VR) headsets; and
4. Leveraging their digital transformation success to improve other areas of the company.

GoldMineCo—Use of EIDI Data in Gold-Mining Operations

GoldMineCo is one of the largest gold-mining companies in the world. The company has mining operations in various continents. In 2018, it produced several million ounces of gold. As of December 31, 2018, the company had more than 50 million ounces of proven and probable gold reserves.

Faced with a downturn in gold prices since 2012 and rising exploration costs, the life cycle to bring new reserves online is extensive with a 10-year minimum to transition from discovery to production. GoldMineCo was under pressure to reduce operational costs and believed that better utilization and analysis of its operations data would identify root causes of production inefficiencies.

The company decided that cost control was paramount, so they set target goals for processing costs. In early 2013, GoldMineCo worked diligently to trim excess costs from its operations, prior to any returns with the PI System. In 2013, 2014, and 2015, they made strides in reducing their all-in-sustaining processing costs. Additional reductions needed to come from innovative use of digital technology.

In 2014, GoldMineCo leadership launched a corporate-wide initiative to standardize KPI reporting for its mines. GoldMineCo's digital transformation team recommended an enterprise-wide real-time data infrastructure approach, acting as the real-time data system of record for consolidated operations and production data. Even though business conditions were difficult and GoldMineCo was in cost-cutting mode, the company decided to standardize on OSIsoft's PI System software via an OSIsoft enterprise agreement (EA). In earlier cases where the PI System was used sporadically throughout the company, they found that individual users achieved cost reductions with the PI System, but these accomplishments were not well documented and not transferred to similar facilities. The company as a whole didn't benefit significantly because of lack of collaboration with corporate personnel. Upon EA execution, PI Systems were deployed across all GoldMineCo sites in a standardized manner, with the expectation that benefits of scale could be achieved with the right employee training and with greater collaboration.

GoldMineCo decided to focus their efforts on the following areas:

- Develop process plant modeling and use PI for root cause analysis for optimization.
- Create a dynamic mass–energy balance.
- Improve water management and reporting.
- Transition to a more effective condition-based maintenance (CBM) program.
- Reduce data gathering time for regulatory compliance reporting.
- Perform big data analytics using PI as the real-time operations data feed.

The company began an extensive user-training initiative and standardized things such as asset data definitions, online calculations, real-time notifications, and reporting. GoldMineCo also decided to build out its PI AF hierarchical data model for several of its sites, which provided a templatized way to define physical assets, real-time calculations, and real-time notification alerts in a reusable manner. This helped their personnel analyze and identify root causes of equipment failures, excess energy usage, and production inefficiencies. These efforts resulted in improved equipment availability and utilization, increased output rates, and reduced dilution, which greatly impacted profitability.

Successful Examples of Enterprise-Wide Digital Transformation 197

Process Plant Optimization Successes at GoldMineCo's Mines and Plants

One of GoldMineCo mines produces more than 1.2 million ounces of gold per year. The double refractory ore contains large amounts of carbon and sulfite, requiring a specialized process that operates within strict environmental parameters. A tiny change in the fluids affects recovery, so the PI System is hugely important in making that work, they reasoned. By using the PI System, GoldMineCo has reduced the total number of environmental deviations by more than 40% and fan trips by over 60%.

At another of their gold recovery facilities, the PI System helped monitor their batch processes that had been reviewed manually, mostly through data forms typed into Excel spreadsheets. Errors were common. Using PI AF and PI Event Frames, reports are now generated with a single click, improving product quality and reducing reporting time by 85%. In addition, different batches can now be easily compared.

At their mine in the western United States, GoldMineCo needed to better understand their energy usage. They could not account for all of the energy usage, because they weren't exactly sure where all the energy was going (underground, open pits, etc.)—they had no granularity at the production level. So GoldMineCo and an integration partner created a static mass–energy balance, to account for the energy purchased each month. The next step was to model the energy flows. Once that was accomplished, GoldMineCo built a PI AF data model to create a dynamic mass–energy balance and to extract and see the data on real-time dashboards. They then ported this dynamic mass–energy balance solution from that mine and the gold recovery site to other GoldMineCo facilities once the necessary instrumentation was put in place.

Water management is a significant operations issue and often a political issue for mining companies, where facilities can be shut down for spills or other water-related compliance issues. Downtime can be very costly. At GoldMineCo sites in South America, the PI System combines real-time data from internal processes along with external data like river rain flow to give the company and its executives the big picture of water consumption. That's important for costs and regulatory compliance.

Combining Time-Series Data and Location Data for Mobile Asset Maintenance

To further reduce operating costs, GoldMineCo has used the PI System to monitor real-time behavior of fixed plant assets (equipment) and its mobile assets. At one of its mines, GoldMineCo has successfully integrated PI System time-series data with Esri's ArcGIS geospatial software, which provides the capability of viewing intelligent maps with real-time and spatial data layered together. This provides GoldMineCo with the ability to monitor their mobile assets spatially and track their performance in real time, which results in lower maintenance costs, reduced downtime, and real-time situational awareness.

GoldMineCo's largest trucks weigh about half a million pounds, stand 30 feet high, and carry 1.2 million pounds fully loaded. These trucks are also equipped with 200–250 sensors, which send the data values to the PI System to predict when failures might occur, sending real-time alerts before a part fails. A blown tire in a typical large mining truck can cost US$60,000.

Smart data layers in Esri's ArcGIS maps can display truck performance data and enable spatial analytics, such as determining if the trucks are reaching their destination in an acceptable amount of time. GoldMineCo built PI System graphic displays showing performance metrics on a mining haul truck. This display can be accessed from tablets, mobile phones, and other devices.

Big Data Analytics—Populated by Site Information

GoldMineCo uses a standard PI System software utility (described in Chapter 10) that publishes contextualized PI System data to business analytic tools, such as Microsoft Power BI. By having access to the production, operations, and energy data from the sites, corporate personnel and executive leadership are able to view role-based and summarized company dashboards for specific use cases, such as energy usage, production costs, and so on.

By leveraging the PI AF templates used at their operating sites, GoldMineCo can easily extract the same data from different sites to populate their Power BI dashboards. A GoldMineCo team member explained, "Once the data is in PI, we can do whatever we want with it."

Other Initiatives to Extract Value from Data

In 2016, the company also decided to improve their gold recovery operations by instituting new methods of identifying and reducing operational hidden losses. To achieve this, they built software templates that were used to capture and segment operational events, to identify when they were running within target ranges versus when they had downtime, maintenance, or production difficulties of some type.

By segregating and classifying these events, they were able to further drill down and assess their operating and energy costs during both normal production periods and during downtime/trouble modes. As a result, they were able to identify significant hidden losses, which were not easily identifiable without consolidated time-series data and the online analytical tools that enabled these valuable discoveries.

This new standardization led to other unexpected benefits, such as reducing the burden on the GoldMineCo IT staff. Because of effective system training, employees were able to analyze and visualize the data themselves.

GoldMineCo's Use of the Digital Plant Template for Metallurgy Analysis

GoldMineCo's innovative use of the *digital plant template* assisted them greatly with improving operational efficiencies, that is, improved equipment availability and utilization, increased tonnages, and reduced dilution. As described in Chapter 4, the digital plant template helps reduce the production variance between scheduled production targets and actual production levels, while reducing operating costs. Consumables are aggregated based on critical operational modes (running OK, trouble, idle, down, and maintenance). Real-time insights are obtained from both EIDI system analytics, cloud-based analytics, such as Microsoft's Power BI, or both. Knowledge workers can access this information anywhere in the corporation.

Using the PI AF capability, GoldMineCo utilized reusable *process unit templates* that were configured with a common set of attributes, for example, process feed rate, water consumption, electricity consumption, and so forth, using standard nomenclature that is quickly and efficiently aggregated and visualized.

GoldMineCo used the digital plant template to help transform plant data into actionable insights for metallurgy analytics. They configured the template with standard analytics and calculations to assess and analyze metrics of common interest, for example, downtime.

A top-down approach was followed whereby the stockpile and crusher, semi-autogenous grinding and ball mill, grinding thickener, carbon-in-leach, carbon-in-columns, and tailings sections were modeled as individual units, with each section modeled using the standard process unit template. The total feed, water, electricity, and reagent consumables were aggregated for each section, and main quality variables and KPI parameters were calculated and monitored. Section operating modes were designated as running OK, trouble, idle, maintenance, or down, based on the section's feed rate and electricity consumption values.

Future Plans

Once GoldMineCo identified impediments to efficient production, they worked on continuously improving their processes and workflows. In 2016, they developed three new capabilities that they are currently deploying:

1. GoldMineCo developed a integrated gold recovery software model, which identifies behavior of the ore type, crushing, grinding, cyanidation gold recovery, and tailing management. This metallurgy analytics initiative is underway using the digital plant template.
2. They optimized their processes through forensic analysis, utilizing predictive analytics such as machine learning.
3. They completed their transition to a proactive CBM program, with real-time notifications generated when equipment performance degrades or exceeds recommended usage. A predictive maintenance program has also been started using IIoT vibration sensors.

The company has plans to further leverage their PI System data to supply real-time data to several of their technology providers in order to diagnose and solve technology and process issues. GoldMineCo also has discussed collecting data from its fleet of drones.

So, Was It Worth It?

As one GoldMineCo stated, "GoldMineCo's digitization initiative completely depends on the PI System to work. We need the data. The data is in PI." He added, "It was money well spent. We have no regrets."

MaterialsCo—Use of EIDI Data in Building Materials Manufacturing

MaterialsCo is a multinational building materials company that manufactures cement and other construction materials worldwide. MaterialsCo employs more than 30,000 people worldwide.

Creating a Single Version of the Truth on a Global Scale

After using PI Systems in numerous cement plants in the late 1990s, MaterialsCo successfully deployed PI Systems in all their cement plants plus several grinding plants worldwide. The initial goal was to provide global visibility into the operations of all facilities. This program was called PIMS (Plant Information Management System). Specifically, MaterialsCo realized they needed to deploy a single unifying platform that integrated and analyzed real-time operations and production data in order to achieve operational intelligence for continuous improvement. They then developed a program that aligned technology initiatives with strategic business goals.

After prioritizing use cases that impact business the most, MaterialsCo decided on the following five initiatives:

1. Provide accurate, reliable information from the dozens of sites to knowledge workers.
2. Enable global standards for process analysis and production reporting.
3. Improve product quality through a standardized methodology to store quality data from many types of legacy laboratory systems.

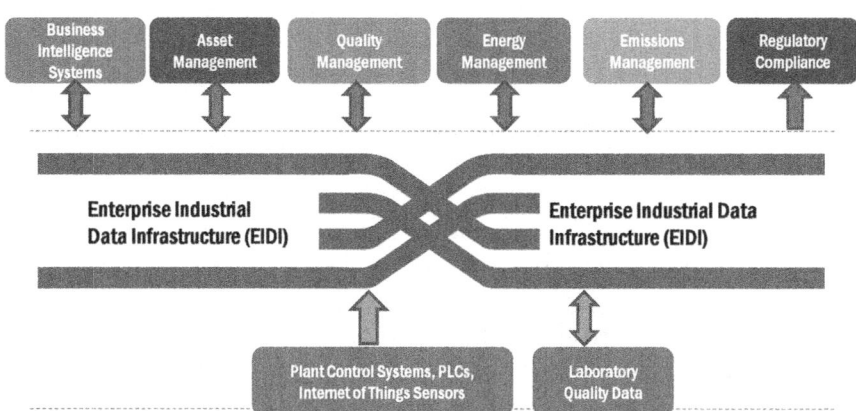

FIGURE 8.4
Typical cement manufacturing company use cases that consume EIDI data for business analytics, quality control, and compliance reporting for purposes of continuous improvement.

4. Reduce energy and maintenance costs through a CBM program to improve asset health.
5. Provide a standard platform for emissions monitoring.

Figure 8.4 illustrates the types of use cases and applications that a typical cement manufacturing company implements.

During the first phase of the process, the MaterialsCo team set out to standardize information across their global organization. They had many plants with different laboratory equipment, control systems, and operating practices. OSIsoft's PI System had been used in several of the plants, but those sites acted independently, using the software to meet its individual needs. Because the PI System had helped these plants achieve their goals, the company made plans to standardize the PI System software as the corporate-wide time-series data infrastructure. This decision was made in order to standardize plant work processes, provide enterprise visibility by their enterprise software integration, and significantly improve collaboration among the MaterialsCo plant sites. It took about three years to deploy the PI System across their cement plants and grinding plants, and then several more years to complete the initial use cases just described.

They started their journey by standardizing things such as data stream naming conventions and online calculations. But they ran into a challenge of providing context to the information so that workers in different roles (operators, production managers, maintenance, finance, planning) could access and utilize the information effectively. The PI System's AF capability provided MaterialsCo the ability to define each cement plant with a similar hierarchical data model. This gave them the following advantages:

- Providing reusable object templates so that MaterialsCo could build the data structure for one cement plant and easily replicate those configurations for all other cement plants.
- Defining common names and common references for plant data point names so they are understood throughout the enterprise.
- Building online performance analytics, totalizers, statistics, and so forth, for one plant and replicating those analytics across all cement plants using the object templates.
- Creating real-time notification alerts based on dynamic equipment or process behavior, so that employees could make quicker decisions and avert serious problems.
- Creating single versions of reports, KPI dashboards, and process displays that can be used by all cement plants through the modular capability of the PI AF software.

Once the PI Systems were deployed across the enterprise, employees began using the information for analysis, reporting, problem solving, and decision-making. As a result, people started to ask more questions and requested additional data for additional analysis. Supporting requests from 70 production sites and regional offices turned out to be a challenging task. So MaterialsCo established a standardized set of production and operations reports. This eliminated the need for individual spreadsheets that duplicated the same purpose, developed by employees moving from one plant to another. By standardizing reports, MaterialsCo eliminated redundant data efforts.

Integration with Corporate Business Intelligence Software

Building upon their success of improving their operations using a global real-time infrastructure for their 70 sites, MaterialsCo focused on integrating PI System data with their enterprise business intelligence systems. To accomplish this, they designed and implemented a standardized method to automatically extract, validate, and transfer data to MaterialsCo's business software suite. This was important because of the following:

- The data needed to be extracted, validated, and sent in the correct format for data ingestion. MaterialsCo did not want to send incorrect data, such as when a sensor was unavailable for a period.
- These data transfers required manual efforts to extracts, cleanse, and publish the data to their corporate business software and other business analytics tools.

MaterialsCo embarked on using newly released OSIsoft-provided software that is designed for extracting and cleansing PI System data and publishing this data to SAP or to the business analytics software of choice. This software eliminated any custom scripting or programming, as users simply configure what they need to transfer. As a result, MaterialsCo significantly reduced this extraction and cleansing effort from one month to five minutes per transfer. This reduction enabled further collaboration because of the visibility and transparency of data.

Forging Ahead—MaterialsCo Production Management Version 2.0

Based on their success with CPM v1.0, MaterialsCo developed the next set of use cases to further mine the operational data to improve production and operations, and better manage their equipment. Initiated in the mid-2010s, the four major goals of CPM v2.0 were to

1. Develop global quality KPIs by consolidating quality and production data on a single platform to monitor overall quality among the cement plants;
2. Further analyze production data through event frame management for root cause analysis of downtimes and to compare downtimes across the fleet of cement plants, specifically, to record, alarm, and analyze all downtimes to minimize operational impacts and prevent future downtimes;
3. Build a cement plant operation management set of standardized dashboards for operations management; and
4. Incorporate vibration analysis system data into the PI System to assist with their CBM program and enhance CBM efforts by sending equipment alerts to their corporate software's maintenance module.

Perhaps the most beneficial part of CPM v2.0 was to start up a centralized control and monitoring center to support the plants by monitoring fleet operations and production and identify possible problems before they became troublesome. This allowed plant personnel to focus on production and letting corporate engineers solve process or equipment issues before the central team notified them.

Lessons Learned from Global Deployment

In building and deploying a real-time infrastructure from their many sites, MaterialsCo recommends the following six best practices:

1. Capture needs and proposals from the operations' network to build the initial list of projects in a collaborative way.
2. Confirm that all proposed initiatives align to company objectives and business impacts.

3. Implement extensive testing programs because testing is always productive and never a waste of time.
4. Build a showroom for testing and demonstration purposes, with the least amount of resources.
5. Document value obtained from deploying each initiative.
6. Continuous improvement is an ongoing cycle: Keep moving forward.

Moving to the Next Level—Artificial Intelligence and Closed-Loop Control

MaterialsCo's latest initiative builds on having used their operating data for descriptive analytics and understanding why things have happened. They are now leveraging their data to develop predictive and prescriptive analytics.

Predictive analytics uses statistical models and calculations to understand and forecast what could happen in the future, such as when specific assets (equipment) will fail. They are also developing prescriptive analytics, in the areas of

- Operations efficiency, forecasting incidents, and downtimes;
- Energy optimization;
- Quality assurance;
- Safety; and
- Alternative fuels substitution.

MaterialsCo is now implementing artificial intelligence (AI)-enabled autonomous control for certain cement plant operations using PI System data with AI tools developed by a software analytics company. Their goals are increased efficiencies leading to lower costs, reduced fuel and energy consumption, better quality, and improved decision-making. The initiative is a phased approach consisting of the following:

- **Predict.** Real-time forecasts improve understanding.
- **Prescribe only.** Prescriptive instructions are validated by plant operators before updating kiln set points. The Autosteer capability is turned off in this mode.
- **Autonomous control.** The Autosteer mode is turned on, with supervised, controlled autopilot operation, integrated with plant control systems.

MaterialsCo has currently achieved the following milestones with this initiative:

Phase I. Forecast predictions in real time. They successfully predicted clinker and air temperatures 15 minutes in advance.

Phase II. Real-time prescriptive recommendations. They relayed recommendations to kiln operators, with model improvements to make recommendations reasonable and actionable.

Phase III. Autosteer operation of the kiln's cooler section. They successfully ran the kiln cooler autonomously, with higher exit air temperatures when Autosteer was engaged.

Phase IV. Multi-site deployment, which is yet to be implemented. In this phase, they will replicate the process using the PI System infrastructure and identify best practices to improve machine learning models.

EIDI Data Use by ChemCo for Specialty Chemical Manufacturing

ChemCo is an international specialty chemical producer of premium polymers, specifically polyurethanes, polycarbonates, and specialty products to customers worldwide. ChemCo employs more than 15,000 people globally. In this section, you will learn how they improved production, reduced energy consumption, and how they go about continuous business and sustainability improvement.

Using Technology to Combat a Decreasing Market

In the twenty-first century, chemical producers have been selling their products to reduced demand and a higher supply for specialty chemical products, with increased competition from Asian companies. In addition, there are several other headwinds. It has become increasingly difficult to use low-cost, geographically convenient feedstock producers, as many countries' manufacturing base has been shrinking, with most of the chemical center clusters now residing in the Middle East, Far East, and Gulf Region. It is also infeasible to change suppliers based on cost because companies cannot transport some of these feedstocks, such as chlorine, over great distances.

Perhaps equally as challenging, in the last decade, many national and international organizations have increased environmental regulations for the chemical industry by a substantial margin. This requires greater emphasis on sustainability and emissions, material handling, and safety issues, which ultimately drives up the cost of manufacturing.

On top of all that, customers are demanding higher quality and more information from their suppliers. ChemCo leadership realizes that they can't just relax and say the entry hurdles to all markets are so high and they'll just keep doing what they have been doing.

The preceding challenges have compelled ChemCo to better utilize digital technology to innovate in three areas:

1. Optimize how ChemCo does its day-to-day business. This entails optimizing chemical processes, keeping plant assets at peak performing conditions, and improving the end-to-end chemical supply chain through enhanced business practices, reducing supply chain costs, and successful interconnectivity of plant OT systems with business IT systems.
2. Augment growth based on how customers do business with ChemCo, developing new strategies and technology to enhance the customer experience.
3. Develop innovative analytics and models using emerging technologies.

Putting Data to Good Use for Improved Operations

To address these challenges, ChemCo has focused on improving operations effectiveness in three areas: safety, cost reduction, and improving quality. Their leadership explains that safety is always their priority. Anything that helps and supports the plants to run safely is appreciated. At the same time, the company needs to reduce costs; one way to get more out of the plants without building new ones is to make sure you get more quality product from the exhausting asset. All this must be accomplished with sustainability in mind, especially with ever-increasing regulations.

In 2007, ChemCo standardized on OSIsoft's PI System to serve as the real-time infrastructure and time-series system of record for their fleet of chemical plants. As of 2016, they had more than 50 PI System servers collecting approximately 1.5 million data points from over 50 production units globally.

Using PI System data, ChemCo developed digital strategies along several tracks:

- Vertical integration within the company, so that key plant information is quickly delivered throughout the enterprise and SMEs can act on that data to make real-time decisions or to discover issues that impede optimized operations
- Business-to-business (B2B) integration with partners and customers, so that information flows quickly and efficiently
- Use of emerging technologies, such as AI analytics and modeling software, to mine and understand patterns in historical production data that are not easily recognized by humans or existing software applications

We will discuss several examples of using data effectively to reduce emissions and to optimize control strategies.

Reducing Emissions and Energy Consumption through Data Analysis

ChemCo uses a three-tiered energy management system. In the first tier, PI System data from all the production plants are used to perform an energy efficiency check. The results are then delivered for analysis to process experts and operators, who identify potential places to reduce energy through small modifications, large investments, or by changing how the plant runs. This provides ChemCo with a clearly prioritized project portfolio that they can then steer investment budgets with and make sure the most promising projects are done.

The second tier is an online monitoring tool for operators. It visually shows, on a minute-by-minute basis, if operators are running within specifications and how energy efficiency can be improved for the specific plant. In the third tier, energy and production data for all plants is standardized, regardless of plant type and process variations. With that standardized information, management can compare results plant by plant and primary energy use per ton of product produced. This provides management and knowledge workers with the transparency they need.

These changes aren't just improving access to information, they're affecting ChemCo's bottom line. In 2009, their program was certified as an energy management system, entitling the company to receive large tax rebates. In addition, ChemCo has reaped the following benefits:

- The company has increased production.
- Energy consumption has been reduced by 30% since 2005 and the company expects to achieve a 50% reduction by 2030.
- Carbon dioxide (CO_2) emissions have been reduced by almost 40% per ton of product produced since 2005, and the company expects to achieve a 50% reduction by 2025.

Asset Management and Other Data-Driven Improvements

ChemCo also has used their PI System data to develop a fleet-wide CBM program designed to do the following:

1. Continuously monitor their plant equipment for sensor malfunctions, such as freezing of data values.
2. Detect offset or drift from optimum operating ranges, caused by conditions such as heat exchanger fouling, excessive vibration, or increased times for normal open/close operations.
3. Totalize the number of times a device turns on or off, which may indicate improper operation or that it has reached the maximum usage time and should be serviced.

4. Synchronize maintenance operations with other software systems, such as their SAP maintenance module.
5. Benchmark similar assets to compare performance of equipment from different suppliers.

Recently, ChemCo has developed new analytic solutions using PI System data that is ingested by a software analytics product that recognizes patterns in large amounts of data. ChemCo used PI System data and analysis tools, together with a third-party analytics software product, to identify and solve several issues:

- **Emissions monitoring.** By identifying when a certain valve opened, there was a pressure drop in off-gas treatment, leading to increased emissions. Once resolved by modifying their control system logic, it reduced off-gas emissions by about 60%.
- **Energy consumption.** By layering five years of energy consumption data (2013–2017) and comparing several specific items, ChemCo was able to reduce energy consumption.
- **Efficient production.** By layering data and comparing good-quality period data with bad-quality period data, ChemCo observed that during the bad-quality periods, one of the flows to the reactor was significantly higher than during good-quality periods. By improving their control system and making sure there would be notifications when this occurred, they significantly reduced bad-quality periods.
- **Batch-to-batch variability.** When producing a specific product, the process went to a hold state when in the reactor. By creating an ideal batch profile or fingerprint from many past batches of a specific product, they were able to identify when batches were deviating from ideal operation. ChemCo made another control system correction that resolved this problem and averted a yearly product loss of over 100 tons.

Lessons Learned and Next Steps

In using real-time and archived plant data for several decades to continuously improve operations, ChemCo recommends the following steps:

1. Leverage existing archived data and look for new ways to analyze information to uncover what is difficult to see or explain. It will turn into business value.
2. Use the information in a company's value chain both horizontally and vertically, integrated with partners and customers.

3. Having a common data infrastructure is an advantage, but a good consistent asset model is needed for fast deployment of initiatives. Big data analysis needs more than just a data lake.
4. Focus on environmental, safety, and sustainability issues. ChemCo has reduced CO_2 emissions by over 35% through PI System data analysis in energy management.
5. Use existing data with emerging analytics software to solve problems by viewing and analyzing the data in new and innovative ways.

Reference

Harclerode, C. 2017. "Data operations transforms fuels value." *PTQ Magazine*, Q1.

9

Beyond the Refinery—Connecting the Ecosystem

We cannot solve problems with the same thinking we used when we created them.

Albert Einstein

Chapter Overview

This chapter describes how ProcIndustries and other manufacturers have transitioned from on-premises software usage for process improvements to near-real-time sharing of their operations data with third-party providers, such as equipment suppliers, process analysis companies, and information technology (IT) systems integrators, in order to more quickly improve their facilities.

This data sharing may be required because of a lack of in-house expertise or in cases where operating companies have service contracts with equipment suppliers that stipulate these providers receive plant performance data for analysis and, in return, provide recommendations for optimal performance. Data sharing is also critical in other scenarios such as contract manufacturing (prevalent in pharmaceutical and life sciences), joint ventures (i.e., upstream oil and gas), and integrated supply chains (i.e., from pit to port). You will follow the ProcIndustries South Texas refinery team as they discover not everything can be resolved with in-house expertise.

We also present two companies who have successfully used a cloud-based data exchange arrangement with third parties, securely transmitting near-real-time data outside their company firewalls. You will learn how Anglo American Platinum, a global diversified mining company and precious metals producer, shared their data with Outotec, a service provider in Finland, through a connected services environment. This service provider is a leading European firm supplying technology and services to companies that are seeking the most sustainable practices in utilizing water and energy.

The second example is a company that provides predictive analytics results to their customers in various industries, in this case, a power generation plant that had surface condenser issues, which led to increased costs. Both companies utilize OSIsoft's PI Cloud Connect services platform, which we explain further.

New Ways of Sharing Data to External Partners for Additional Value

For the past few decades, companies like ProcIndustries have deployed on-premises OSIsoft PI System software as enterprise industrial data infrastructure (EIDI) systems. These EIDI systems are installed at operating plants or control centers, with subsets of this contextualized data continuously transferred internally to other EIDI/PI Systems located within the enterprise, behind the company's business network firewall. There are several ways to architect this, but it suffices to say, this data is generally not available for public consumption.

In the 2010s, it became more common for companies to be able to share subsets of their data with trusted partners in a real-time or near-real-time environment. Using a simple example, a turbine manufacturer and a power generation plant enter into a long-term service agreement. Per the terms of their agreement, the turbine supplier needs some of the plant's turbine operating data to

- Determine if the turbine is being run correctly within recommended rates, as specified by the turbine manufacturer;
- Diagnose unexpected operating problems;
- Suggest repairs that should occur sooner than scheduled maintenance times; and
- Determine if the turbine can withstand higher rates of production during peak periods.

There are various reasons companies outsource plant operations or equipment maintenance to third parties. Sometimes the entire plant operation is outsourced. In other cases, an enterprise does not have sufficient in-house expertise to analyze data for a specific use case. Or the equipment provider requires the operating company supply subsets of data in an easily consumable fashion. These requirements have evolved to a real-time or near-real-time feed via an Internet or cloud-based mechanism.

As a result, new technology has emerged to share increasing amounts of data among disparate organizations, transmitting data accurately and securely outside the company's firewalls to trusted third parties as needed. Specifically, OSIsoft developed a cloud-based subscription platform as a service (PaaS) called PI Cloud Connect or PI CC. PI CC simplifies third-party access via secure transmission of plant information to subscribers who are authorized by the owner of the data, who is usually the manufacturing company. The owner simply selects who is to receive the data and publishes the selected data points to that subscriber. Data exchange is accomplished via the cloud, utilizing best cybersecurity practices. In the following examples, PI CC uses Microsoft's Azure cloud platform services to facilitate bilateral data exchange.

Combining these technologies enables remote monitoring use cases that significantly improve operational effectiveness, resulting in lower production costs. Internal and external experts become partners to analyze problems and abnormal situations to determine ways to improve and optimize operations, without physical travel to the site.

Circling back to the ProcIndustries South Texas refinery, the digital transformation team takes pride in the fact that they have used the EIDI to analyze their refinery data and have been able to resolve nagging issues that impeded production goals, such as

- Identify which factors lead to increased yields;
- Spot heat-exchanger corrosion, leakage, and fouling;
- Locate excessive or low vapor flow in the distillation columns;
- Locate leaky valves and tube leaks; and
- Identify equipment that was approaching imminent failure.

By using templated OSIsoft PI Asset Framework (AF) online analytics and notifications, the team was able to be notified in time to act before a catastrophe occurred. When a situation wasn't time critical, they leveraged the experience and expertise of their engineers to recommend solutions. Although they are pleased with their progress toward a digital transformation, the team realizes that there are some specialized process and equipment issues that are beyond their capabilities. To identify which unsolved or complex problems can best be solved more quickly by one of their suppliers or technology partners who possesses deeper domain knowledge, the team holds a brainstorming session. They prioritize three use cases that can be best solved externally:

1. **Catalyst performance optimization.** Over time, the team has observed that something is causing decreased reformate yields. They believe their catalyst provider can perform offline analysis using the reformer's archived historical data and recommend optimum settings.
2. **Data exchange with their process technology provider.** The team surmises that the provider who licenses alkylation and isomerization process technology to the refinery has the ability to monitor

refinery performance and determine how ProcIndustries is performing relative to expected results. As a result, the technology provider can offer recommendations for performance improvement. Perhaps ongoing oversight is possible as well.
3. **More in-depth analysis of their compressor performance.** By sending real-time data from critical equipment, such as fluid catalytic cracking unit (FCCU) wet gas compressors and air blowers to the equipment provider, the equipment provider can completely analyze the equipment and offer the team solutions.

After selecting the appropriate third parties, the first task is to determine exactly what data should be sent to each of the third parties. The team tasked one of the refinery engineers to provide a list of data points for each provider and time durations for each data set. The engineer suggested setting up PI event frames to represent optimal time periods for analysis. In this way, the data is likely to give an accurate portrayal of the problem issues. The engineer also recommended using the PI AF asset model database, whereby data from the assets in question can be sent in a logical way so that the solution provider can understand the context of each data value. It is likely that the provider already has an AF database for their asset analytics that can be easily tailored for the South Texas refinery.

The next order of business was to work with ProcIndustries IT management to get approval for and set up the data communications connection with the service or equipment provider to transmit the real-time refinery data. Peter Argus (continuous-improvement manager) and Bill Roberts (vice president of operations) scheduled a meeting with ProcIndustries' IT manager, Pat Verlaine, to discuss the following three items:

1. An overview of the business needs and reason for outsourcing
2. A summary of the data that will be transmitted
3. How the data exchange connection (PI CC) will be deployed and maintained to ensure a secure connection, protecting the ProcIndustries IT environment as well as the information being exchanged

Concurrently, Bill has identified and evaluated providers that use this emerging technology with their customers, in which customers securely transmit real-time operating data to providers via the cloud. These providers ensure that the manufacturing company's information will not be shared or compromised.

Bill also learns that several of their service and equipment providers are using customer data exchange platforms that utilize the same EIDI software system (OSIsoft's PI System) as the backbone of their remote analysis service (PI CC).

PI Cloud Connect

The digital transformation team meets with Pat and they discuss how the PI CC service can enable easier and quicker resolution of problems that are better solved by ProcIndustries' partners. Pat agrees to evaluate the PI CC PaaS to see if it meets the company's requirements, which are as follows:

- ProcIndustries defines the data transmitted and controls third-party access to the data.
- No external entity can access data without ProcIndustries physically authorizing it.
- The connection must be 99.9% available and accurately transmit the data.
- The connection must be 100% secure so that neither the ProcIndustries IT environment nor the ProcIndustries data are compromised.
- The third party cannot have direct access to the South Texas refinery EIDI, except to send analysis results to the EIDI on a network that cannot access the control network or key business information.
- The platform should require minimal effort to use, with software deployment and maintenance activities mostly performed by the software vendor, including managing the cloud hosting environment.
- The platform needs to be scalable and adaptable to typical IT cloud environments (Microsoft Azure or Amazon Web Services), without excessive deployment efforts.
- There should not be significant translation or cross-mapping issues between data points.

Peter explains and summarizes what he learned about PI CC from attending the OSIsoft annual user conference:

- It is architected using secured data sharing without virtual private networks (VPNs).
- Early adapter companies related that it was easy to install, connect, and configure.
- The software is managed by OSIsoft with a minimal on-premises footprint.
- It's a scalable solution based on the Microsoft Azure cloud environment.
- It utilizes a web-based portal for configuration and monitoring.
- Because the EIDI has the capability to store future data, such as model predictions, our people can track real-time variance between what the provider has estimated and our actual results.

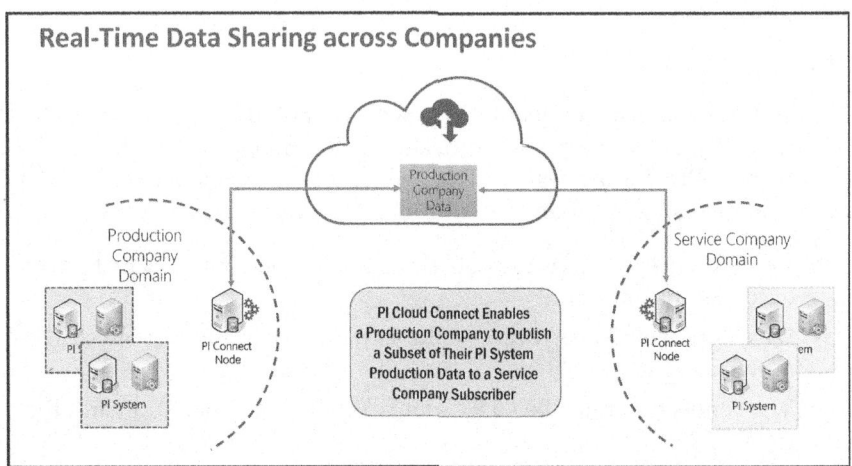

FIGURE 9.1
PI Cloud Connect architecture for production companies and service providers.

Figure 9.1 shows the interconnection between a customer and a service provider using PI CC. Once a connection is enabled in PI CC, the publisher of the data (typically the manufacturing company) browses their PI AF asset object model database and selects what data they wish to share with the subscriber, generally the equipment or service provider.

The exchange of data in the PI CC utilizes Microsoft Azure Cloud Services. The data published can include real-time streaming data as well as any metadata stored in the data model. The manufacturer can optionally publish the complete asset object model via PI CC. Once a copy of the asset object model is received by the third-party provider, the actual data is published and begins streaming in real time. Peter offers to provide references that show successful OSIsoft customer implementations.

Figure 9.2 shows several examples of entities that use a connected services environment to assist their customers with operational challenges, using a "just-in-time" model for analyzing customer data for various use cases.

The four circles represent the senders and recipients of the real-time data shared via the cloud. The left circle represents the production sites that would normally send a subset of the actual production data. For example, they might designate PI CC to send the subset of their data relating to compressor performance to their compressor supplier, who subscribes to receive that data. The top circle could represent the company's headquarters, whose business management receives aggregated production rates, costs, cycle times, cost of energy and water consumables, key performance indicators (KPIs), product quality, and so on. Business leadership likely would not want detailed equipment performance data.

The right circle shows a company that might want to provide limited information to joint partners, perhaps a production plant that is owned by both

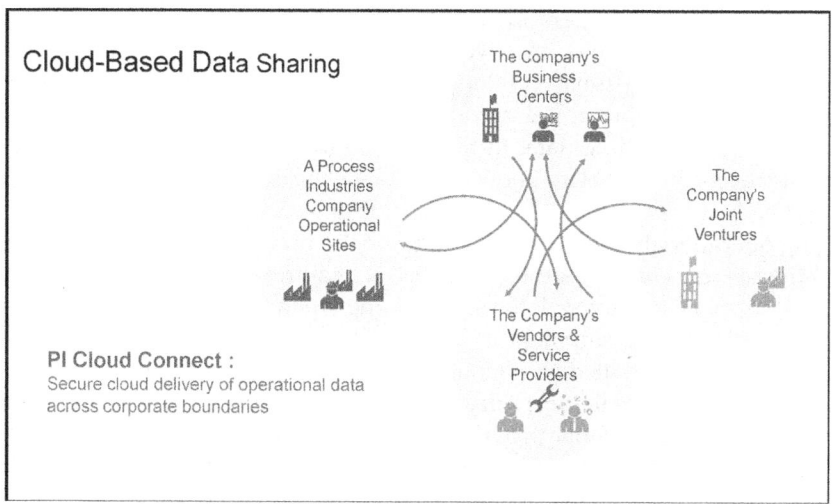

FIGURE 9.2
Data sharing using connected services for process manufacturers to their business partners and service providers.

entities, while the remaining plants are solely owned and no information would be shared with the partner. The bottom circle shows multiple entities that can utilize a direct data connection between the production company owning the assets and their equipment vendors, service providers, and domain subject-matter experts. These companies need visibility into specific subsets of the plant data and may not have direct visibility into the manufacturing operations data. The production or operations data transported through the cloud makes serving their customers quicker and more efficient.

PI Cloud Connect Customer Examples

Peter recounts the PI CC customer presentations he viewed at the most recent OSIsoft user conference. The first use case describes how the mining and precious metals producer Anglo American Platinum solicited assistance from their equipment and service provider Outotec to help them optimize their filter performance. The second case describes a large specialty chemicals and solutions provider who utilized the PI CC platform to provide condenser analytics and recommendations to their power generation customers.

Anglo American Platinum—Outotec

Anglo American Platinum (AAP) strived to optimize energy consumption and reduce overall operating costs. They explored receiving assistance from external service providers in these areas. AAP made the decision to outsource the analysis and improvement of these issues by providing operational data to Finnish-based filter provider and engineering consulting firm Outotec. AAP's main concern was to find a reliable and secure method of sharing their real-time data. In the past, large data sets were exchanged in spreadsheets with no clear knowledge of how the data was obtained. Usually, the data had been filtered and averaged, losing its value to build models for using operational modes in the analysis of the process units and equipment.

Mineral concentrates are the product of ore-dressing operations whereby valuable metals recovered through mining operations are separated from waste rock prior to shipment to market. In many mining operations, ore is crushed and wet milled to liberate the valuable mineral. This concentrated slurry is then filtered to form a dry mineral concentrate that is shipped to refineries that produce metallic products. The type of filtration equipment required depends on the particle size, mineralogy, and shipping requirements. As with all mining operations, the required equipment should be robust and reliable even under the toughest operating conditions (Bascur and Halhead 2013).

AAP uses many filters in their operations. Outotec offered two series of automatic pressure filters: the Larox PF and the Larox DS. Outotec had been providing support by having an engineer periodically visit AAP to check the filters. However, AAP preferred 24/7 support. To accomplish this, Outotec required the filter data, which was provided via PI CC from the EIDI implemented by AAP (Bascur et al. 2016).

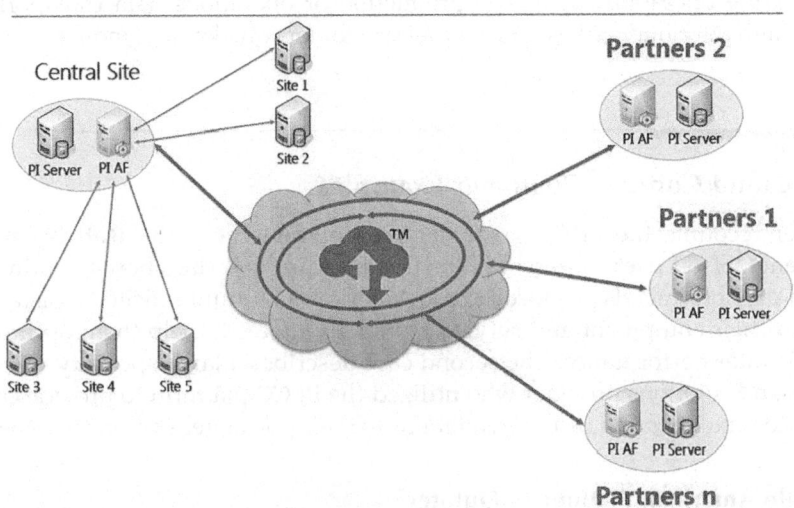

FIGURE 9.3
PI Cloud Connect environment at Anglo American Platinum.

The EIDI selectively shared using PI CC between AAP and Outotec (Halhead 2014; Figure 9.3). The data published by AAP represented sets of advanced filters used for material separation. As such, the operational data was shared by the customer to the service provider in real time. It drastically reduced the time spent in accessing the data and enhanced its use by Outotec. As a result, Outotec advised optimal tuning of the filter cloth and cycling times to reduce humidity of the concentrates, resulting in optimal energy use. AAP continues to explore new possible use cases as they are now capable of reliably and safely transmitting their data at the original resolution to service providers.

The requirements for AAP and Outotec to create a reliable, secure data exchange were

- Easy transmittal of data by AAP and easy data access by Outotec;
- A secure connection that did not compromise AAP IT network entry;
- Data exchanges that were auditable; and
- A standard, time-efficient method to publish new data points for Outotec's consumption.

In Figure 9.4, AAP's EIDI AF data model is shared by the plant to the service provider, so the provider has much better insight into the data being collected and more information about the particular asset they are monitoring. Optimizing the filter operations has several benefits for both AAP and Outotec.

Benefits for Anglo American Platinum

- Reduced energy consumption by controlling humidity for the concentrate during smelting operations

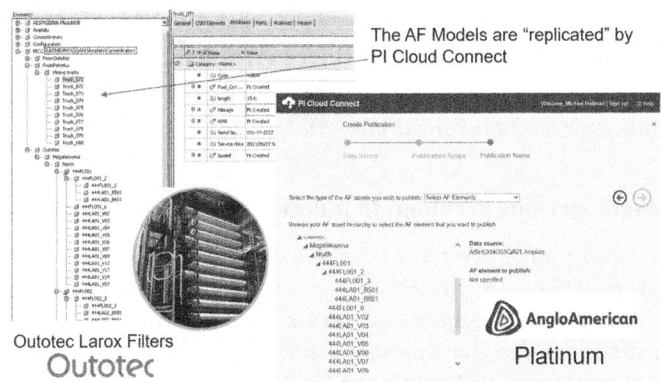

FIGURE 9.4
AAP's asset data hierarchy is made available to Outotec for remote analysis.

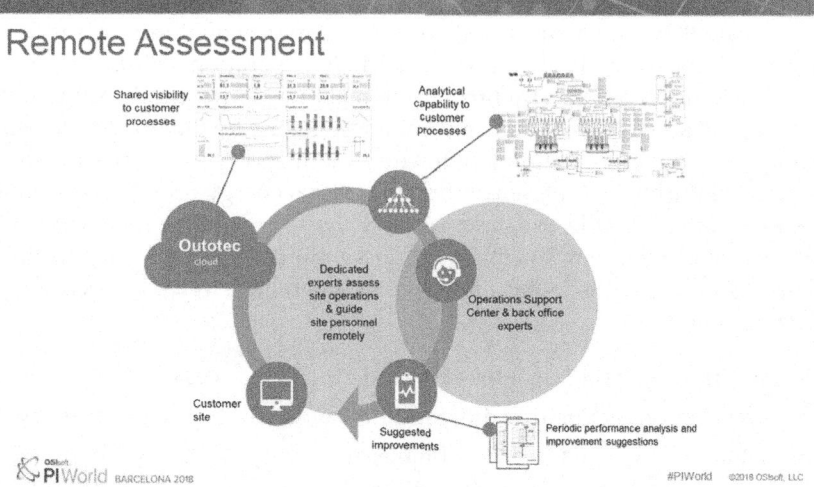

FIGURE 9.5
Outotec remote service methodology and process workflow.

- Reduced usage of concentrate water, which in turn decreased transport costs, because of reduced material weight
- Optimized operations caused by reduced cycle times
- Optimal filter cycle times and proactive equipment monitoring resulting in increased filter availability (Figure 9.5)

Benefits for Outotec

- Provided a reliable, constant 24/7 remote service to their customers (Figure 9.6)
- Reduced travel time and travel expenses for their customers
- Enhanced their knowledge of filter performance
- Utilized highly granular filter data, at their original resolution, versus averaged and summarized data, which is much less useful

Cloud-Based Services Provided to a Power Generation Company

One of the companies we have worked with is a leader in specialty products, best practices in sustainable process technology and equipment hygiene, and energy technology services. They provide chemistry programs and services to the upstream and midstream oil and gas industry, downstream refinery and petrochemical plants, power generation operations, and many other use cases. Through onsite problem solving and the application of innovative technologies, such as cloud-based predictive

Beyond the Refinery—Connecting the Ecosystem

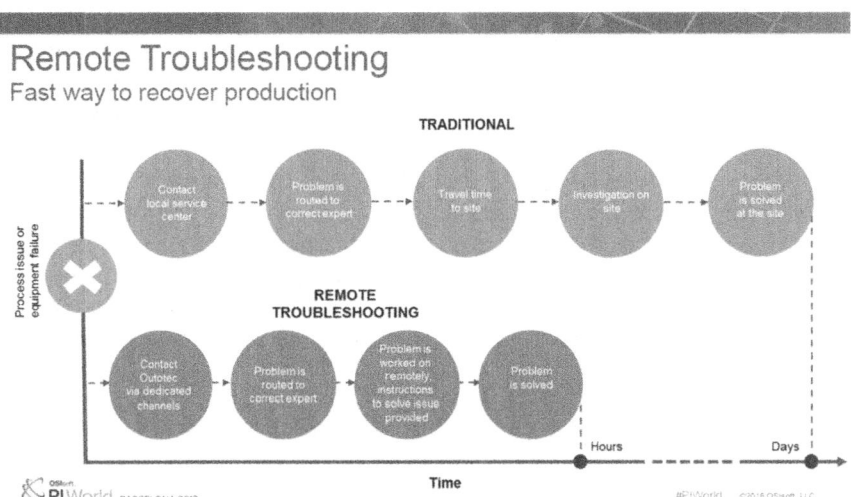

FIGURE 9.6
Traditional versus remote troubleshooting scenarios. (From O'Sullivan, K., "Outotec's digital services platform: How OSIsoft is helping transform our business." Presented at the 2015 EMEA Users Conference, Prague. www.osisoft.com/Templates/item-abstract.aspx?id=12792, 2015.)

analytics, this company strives to help their customers maximize production, optimize water usage, and overcome technical challenges.

In the power generation industry, they monitor customer condenser and heat-exchanger activity 24/7 in their cloud service activity center. They have collected data from more than 200 condensers at 100 customer sites in the United States. Relying on their many decades of experience, they have imparted their domain knowledge into digitized rules and key performance metrics, resulting in the creation of their predictive analytic models that are run at the service provider's centers.

In the past, customers sent this service provider data sets through Excel files and other customized methods. Now they have adopted the PI CC platform, as 80% of their power generation customers utilize OSIsoft's PI System as their plant EIDI.

This service provider has also developed PI AF database templates for condensers and other types of plant equipment. With that, they are able to create a "digital twin" representation of the customer's physical environment.

When the analytical models are run and compared against the customer's actual data, the servicing company calculates equipment performance versus design performance and can determine if there is a problem with current operations. It may be that the condenser has tube leakage or a fouling issue, or there are cooling water cleanliness issues. These conditions may cause high condenser backpressure levels, resulting in increased energy costs and potential equipment failure. In some cases, they have predicted potential failures or problems more than three months before actual failure. As a guideline, the service provider asserts that 1 inch of increased condenser backpressure equals a 1% increase in energy use for combined cycle plants. Solving this can result in hundreds of thousands of dollars per year in fuel cost savings.

To enable this 24/7 service, they partnered with a large systems integrator and thought leader for analytics and visualization support, Microsoft (Azure Cloud Services), and OSIsoft (PI CC). Because of their cloud-based analytics service, they have helped customers realize significant operating and energy cost savings, predicted performance of critical assets, prevented unscheduled downtime and lost production, and prolonged asset life and production run times. As an added bonus, they provide both site-level and enterprise-level KPI dashboards showing all condenser KPIs and color-coded operating status.

PI Cloud Connect Scalability

One of the advantages of the PI CC service is its scalability on both the customer and the service provider sides, meaning that a third-party service provider or equipment manufacturer can receive data from many sources, either from multiple ProcIndustries facilities or from facilities owned by multiple companies. Figure 9.7 shows a representation of this scalability.

With the proliferation of industrial Internet of things (IIoT) sensors and devices being deployed at rapid rates, there exists a need for this remote

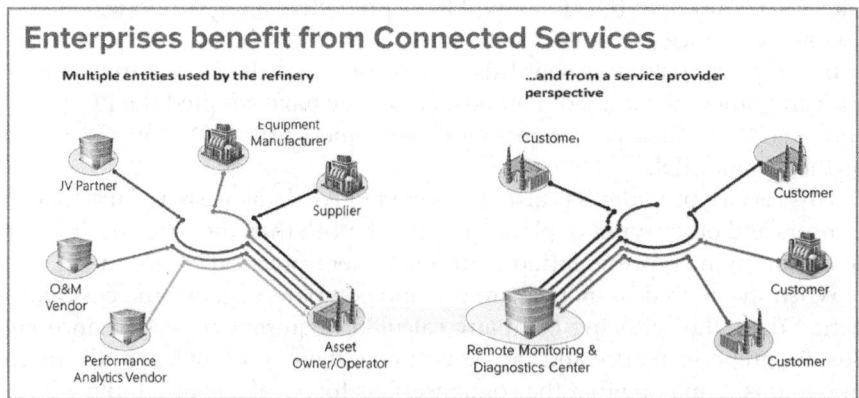

FIGURE 9.7
Connected services community environment where a vendor receives real-time data concurrently from multiple customers using the PI Cloud Connect software, or a manufacturer uses different providers.

servicing model. It will be embraced by IIoT equipment manufacturers and domain experts that utilize this equipment for their applications because software companies like OSIsoft have developed a standard message formatting as one method of having these IIoT sensors or devices easily send data to providers.

The top reasons for transformation through connected services for providers are

- Enhanced revenue growth through 24/7 remote monitoring and diagnostic services;
- Increased efficiency because no travel to the site is required;
- Market differentiation as a remote service provider;
- Improvements in equipment reliability and performance;
- Enabling advanced analytics, for example, machine learning and artificial intelligence tools; and
- Highly granular and streaming customer data for improved analysis and to remotely create a digital twin, which represents the customer's physical environment.

What You Should Take Away

Remote servicing using real-time streaming customer data has emerged as an effective strategy for equipment vendors, process technology providers, or other service companies to better serve their customers. Connected services platforms represent one of the key digital transformation strategies that are aiding process industry and energy producers to become more effective and efficient. Via secure and reliable cloud services, IT departments are able to successfully exchange key operating data between customers and suppliers.

The OSIsoft PI Cloud Connect platform, used by the companies in this chapter, facilitates secure and reliable real-time data exchanges between operating companies and service providers without outside access to the operating company's IT environment. With the EIDI (PI System) capability to store and display future data, end users can display real-time variance between vendor predictions and what actually occurs at the plant.

It enables services providers to build their own digital plant templates to solve common problems for their customers around the world. They will likely provide additional services that cannot be presently imagined. This is one of the most powerful platform business models to improve asset optimization, energy conservation, and yield maximization.

References

Bascur, O.A., and Halhead, M. 2013. "Enterprise manufacturing services to enhance energy effectiveness and sustainability management at Anglo Plats." Presented at the MMM 2013 International Symposium, International Federation of Automatic Control (IFAC), August 24–27, San Diego, CA.

Bascur, O.A., Halhead, M., Garrigues, L., and Jarvis, M. 2016. "Mineral processing plant asset and energy optimization: The calming cloud over operations." In *International Mineral Processing Congress (IMPC XXVIII) Proceedings*. Quebec City, Canada: IMPC.

Halhead, M. 2014. "Calming cloud over the supply chain." Presented at the OSIsoft Regional Seminar, Johannesburg, South Africa.

O'Sullivan, K. 2015. "Outotec's digital services platform: How OSIsoft is helping transform our business." Presented at the 2015 EMEA Users Conference, Prague. www.osisoft.com/Templates/item-abstract.aspx?id=12792.

Additional Reading

Hagiu, A., and Wright, J. 2015. "Multi-sided platforms." Working paper, Harvard Business School, Cambridge, MA.

O'Sullivan, K., and Niedoba, P. 2016. "Remote services enabled by IIOT: A business transformation opportunity for equipment vendors and services providers." Webinar (March 9), OSIsoft, San Leandro, CA.

Rogers, D.L. 2016. *The Digital Playbook*. New York: Columbia University Press.

10

Operational and Business Analytics Integration

The successful man is the one who finds out what is the matter with his business before his competitors do.

Roy L. Smith

Chapter Overview

In this chapter, the digital transformation team investigates how to achieve additional insight from the real-time operations and production data they have been collecting at the South Texas refinery. They have successfully obtained results using the enterprise industrial data infrastructure (EIDI)'s online analytics. Now, they speculate that additional benefits can be achieved using offline business analytics and predictive analytic tools, simply by leveraging their existing information. Figure 10.1 illustrates some of the topics that larger scale analytics are able to address.

In working with the corporate information technology (IT) department, the team learns how to take advantage of a standard EIDI tool that helps them extract and contextualize their data, and then transmit it in a format that makes it easy for big data analytics to ingest. We also discuss several approaches for data input to offline analytics: using an EIDI to contextualize data, using a data lake, and using both.

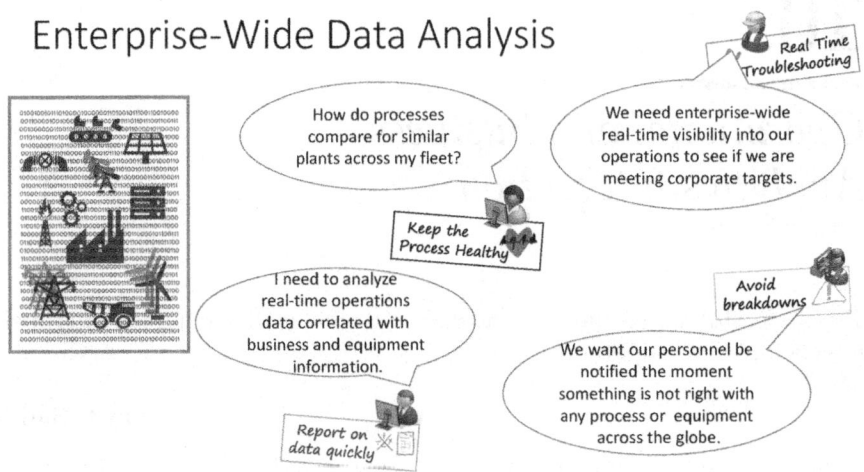

FIGURE 10.1
Advanced analytics solves broader questions.

From Operational Intelligence to Enterprise Intelligence

In process plants and refineries, time-series databases are most prevalent for operations, maintenance, and engineering use cases, with supporting relational databases. Time-series data is used for finding root causes of plant problems, production reporting, calculating key performance indicators (KPIs), and so on. Conversely, business databases are more transactional, more interrelated, and house data such as production costs. If the enterprise wishes to more clearly understand and make impactful decisions regarding total production costs, time-series operations data is combined with business data for greater insights.

In previous chapters, we illustrated various ways a time-series based enterprise industrial data infrastructure (EIDI) can positively impact an enterprise. For example, a maintenance engineer analyzes a turbine's behavior to avoid an unscheduled shutdown. Subsequently, the engineer investigates and resolves the issue quickly to minimize downtime. A business analyst may study plant efficiency from production results and throughput. Plant operators make sure that today's plant production is complying with scheduled targets.

Operational and Business Analytics Integration 227

Types of Advanced Analytics

Before we delve into advanced corporate analytics, let's categorize the types of analytics that production companies have at their disposal. They are listed from the least to most complex, and categorized as OT (operational technology/ engineering) analytics, as IT analytics, or both.

1. **Online EIDI analytics (OT).** Some examples are performance calculations, totalized values, statistics, soft sensors, or whether an early warning condition exists. These are Excel-like calculations that are configured and executed in the EIDI system. They are processed when input data is received, so they are time deterministic. They generally are the easiest to configure and deploy because they are part of the EIDI. Most of these calculations use EIDI data as input.

2. **Statistical quality control (SQC) charts (OT).** These charts are populated by EIDI data, assuming the specific EIDI system supports this feature. The user simply has to configure the desired SQC chart type and the configuration parameters for that specific chart. Examples of configuration parameters are upper and lower control limits or specifying alarm rules. For example, when x number of values fall into a particular zone or are above/below a limit line, this constitutes an SQC alarm. SQC alarms are typically used for monitoring product quality, but they can be used for monitoring other inputs, such as soft sensors. Some EIDI systems have real-time SQC capability, where SQC alarms are automatically calculated, without the user having to view the charts manually.

3. **Microsoft Excel (OT or IT).** Most people at plant facilities are very comfortable with using Excel spreadsheets for various applications. Financial people prefer Excel over traditional EIDI displays and trends. Engineers can import EIDI data into an Excel spreadsheet, where they analyze groups of data variables, get statistical information on historical data, visualize the data in Excel-provided charts, or expose the data to Excel's many mathematical formulas. Software vendors often include some of their software as an add-in menu to Excel, where it can use spreadsheet data to perform calculations, such as linear regression, or to model the data.

4. **MATLAB and software languages such as R and Python (IT).** These are powerful and often-used tools for process data analytics. These tools have libraries that contain powerful algorithms for

advanced calculations that can be used for all layers of process data analytics, including descriptive, diagnostic, predictive, and prescriptive analytics. The tools help turn process data into meaningful insights by applying techniques such as machine learning. As an example, these analytics can be used to identify process or equipment behavior and make patterns visible. These analytics can predict expected performance. The EIDI provides its operational data as input, typically highly granular historical process and asset data. It may even be filtered by context, such as a specific product grade or during a certain section of the manufacturing process. For machine-learning analytics, the predicted results may be sent back to the EIDI system to assist operations personnel in making better decisions.

5. **Process or equipment models and simulations (OT).** These tools are typically used to simulate either process or equipment behavior under given conditions. Process models can be based on engineering principles, either static or dynamic, linear or non-linear. Alternatively, they can be agnostic to the physical process or environment and rely solely on the historical data they consume to make decisions and predict future behavior. These predictive tools are important to effective operations and have been used for decades. Some of these tools also provide closed-loop or advisory (open-loop) control of a process. Many companies are creating digital twins of the plant or refinery, which is used for analytics and software development. Later in the chapter we discuss digital twins.

6. **Offline and big data analytics (IT).** These are the newest and most complex analytics. They may use operations and process data for their inputs and often use multiple sources of data (e.g., operational, quality, financial, and messaging) as input. Some examples of these are advanced visualization tools (e.g., Tableau, Spotfire, Power BI, Qlik) or modeling/machine-learning based tools, such as Azure's Machine Learning Studio. Companies may also develop their own proprietary analytics or use a combination of them. The analytics may reside inside the company's corporate firewall, or they may be cloud-based, such as Microsoft's Azure or Amazon Web Services' suite of products. Because they are not connected to an EIDI system, they require either published data sets of contextualized EIDI data, or input may come as raw or prepared data sets from a data lake.

Holistic Analytics to Meet Business Objectives

In cases when refineries, plants, or business units are utilizing an EIDI in their operating plants, it may make sense to analyze problems holistically, to determine if improvement is possible across the entire refinery, a fleet of refineries, or for the business unit. To do that, companies often combine contextualized production and operations data with business data to identify roadblocks that keep the facility from operating more effectively. Big data analytics enable management to effectively plan where and when to allocate funding for debottlenecking purposes, expand physical capacity, or when to optimally schedule maintenance outages.

In addition, artificial intelligence and predictive analytics provide estimates on how long assets will operate effectively or when they need more production capacity. For example, during the oil and gas industry downturn of the mid 2010s, the larger oil and gas operating companies performed large-scale analyses to identify controllable losses, production obstacles, and correctable equipment inefficiencies to optimize their upstream, midstream, and downstream operations. Data analysis of this scale usually requires a large quantity of high-fidelity time-series data combined with other business and equipment information to feed process and equipment models, big data analytics, and predictive analytics.

When companies undertake such a large data analysis project, specific IT and data science skills are required. As such, when using plant and engineering information, the OT world converges with the IT world. To implement a project of this magnitude successfully, they must align and work together, bringing their individual disciplines and skill sets to the table. An example of an analytic integrating multiple types of data is shown in Figure 10.2.

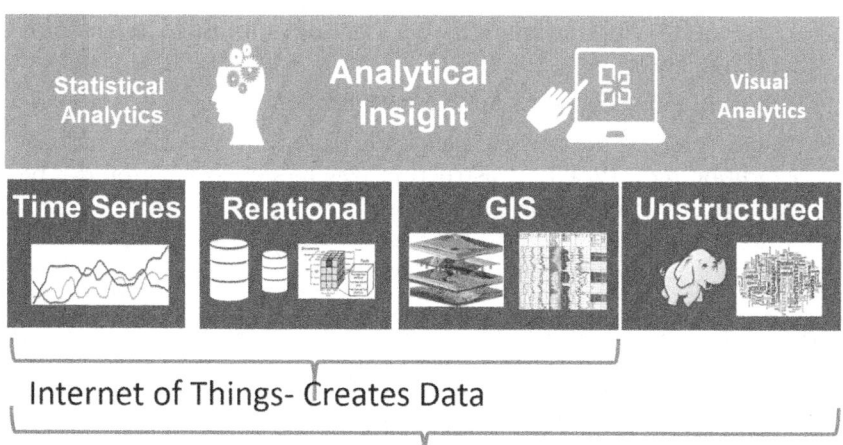

FIGURE 10.2
Data types integrated for big data analytics.

Plant and business operations personnel bring extensive experience and knowledge of how plants and processes operate. They are generally aware what types of problems exist, but may not know exactly where to find them. They are very familiar with time-series data and know how to utilize it. More than likely, they can determine which data values are accurate and which values are outliers that should be discarded and not be used by the analytics. They may even know how to incorporate cost data into operations and production data, but likely do not know where these costs reside or know how to access this data in the IT world.

IT workers and data scientists possess skills that complement the business people and engineers. While they may not have the technical knowledge of how product is made and which issues limit plant effectiveness, they can determine which types of data best facilitate big data analytics and how best to utilize them. Some examples are financial databases, equipment databases, and geospatial databases.

Relational databases are more transactional than real-time databases and designed to easily access relationships between data. If used, geospatial databases are optimized for spatial interactions among data. Similarly, operations data needs to be organized and formatted for IT analytics. Fortunately, this operations data can be extracted and converted to a format that the business analytic system or geographic system can easily ingest to link process data to non-time-based systems or databases to solve business issues, such as the following:

- Should we add capacity at a particular location?
- Where are the hidden production losses?
- When will a specific asset be expected to fail?
- What are the costs of maintaining existing equipment versus purchasing new assets?
- Where are the best performing assets located?
- What would be the energy costs if the heating, ventilation, and air-conditioning (HVAC) infrastructure was modernized? Would it be cost-effective in the long run to do so?

ProcIndustries Integrates Operational and Business Data

The ProcIndustries team, led by Peter Argus (continuous-improvement manager) looks forward to working with corporate IT director Pat Verlaine to start implementing business analytics and identifying artificial intelligence

Operational and Business Analytics Integration 231

tools that leverage the EIDI data they already possess. Peter's vision is clear. He would like to use the collected refinery data to accomplish the following:

- Improve equipment uptime through data analysis. If possible, determine the expected lifetime of critical refinery assets.
- Avoid product quality excursions.
- Increase product yields.
- Increase production flow from raw materials, during process transformations, and shipping final product to customers.
- Minimize consumable resources (e.g., power, water, steam, hydrogen) while maintaining acceptable yield, quality, and throughput.
- Minimize environmental and safety issues.
- Simplify production, quality, equipment, and safety and environmental reporting.

Time-series data enables companies to use continuously fed real-time sensor data with derived variables and event-framed data (see Chapter 3). A time-series derived data variable can measure standard deviation, maximum, minimum, average, totalized value, mode, and other things. These derived variables can also be time-sliced to frame and capture data within a specific production or operational event. The EIDI provides this rich set of information in real time, which assists when implementing optimization strategies, such as machine learning (ML) for predictive asset behavior. These augmented data sets become available to train predictive analytics models using machine-learning tools. In turn, these models train algorithms with added information about variability in the data. The basic raw sensor data is often insufficient to accomplish advanced analytics.

The availability of the data enables these algorithms to learn and continuously improve, and to adapt to process changes as production or equipment behavior deteriorates. Having the historical data available enables forward-looking models. By extracting contextual historical data and transforming the data into a data set that the algorithms are easily able to consume as input, predictions regarding process unit behavior are made. As a result, users and process control systems are able to make better-informed decisions that avoid production losses; inferior quality products; and reduce wasted energy, water, and steam.

Time-series data can be used to generate real-time alerts indicating impending equipment failures. Systems are configured to prescribe detailed actions that mitigate or solve problems before an unscheduled downtime event occurs (see Chapter 6). Peter Argus commented, "Today we can have both production prediction models and asset equipment warnings in real time, using much larger data sets than ever before. Predictive models can recognize precise patterns that indicate degradation and impending failure." Peter added that their EIDI had the ability to store and display future time data, meaning refinery personnel can monitor variances in real time. They can observe how equipment or a process behaves, measured against predicted values.

Chuck Smith (instrumentation and process control engineer) emphasized, "These data patterns are the key to developing predictive analytics models." In the past, ProcIndustries used mathematical/statistical models based on engineering principles. They are still a very important basis for the development of these models, but the inability to recognize time patterns embedded in the operational data makes them less attractive.

Peter pointed out that they need a standard, simple way to extract the data for the increasing number of queries people and software systems will need, once people gain access to this information: "The team has to investigate how to develop a standardized tool that reduces the time for data extraction. The operations teams will be more engaged if the extraction time is short and they don't have to develop custom scripts."

Integration to Corporate Analytics Systems

The time-series and event-framed data available in the EIDI is extracted to machine tools, business intelligence tools, or a data lake as shown in Figure 10.3. The team has been looking at business tools, such as MATLAB,

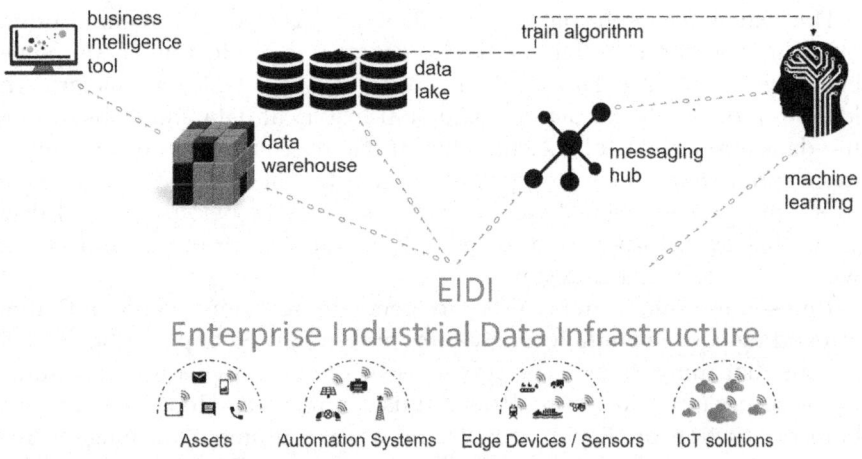

FIGURE 10.3
Integrating EIDI time-series data with big data analytics.

Python, Tableau, Microsoft's Power BI, and Amazon's QuickSight. They also use Microsoft's Azure Machine Learning and Amazon Machine Learning. They have worked on a prototype using Microsoft's analysis tools via EIDI Open Database Connectivity (ODBC). Their EIDI offers a standard integration tool that simplifies integration with offline analytics tools.

Using a Real-Time Data Infrastructure versus a Data Lake Approach

Many IT departments of major energy, process, and manufacturing companies use a real-time data infrastructure or EIDI at the plant operations and engineering levels. Typically, the sensor data, either control system–based or separate Internet of things (IoT) sensors, are stored at its original resolution (excluding a noise filter) into a permanent historical data archive. Additionally, the sensor data is often transformed into more valuable inferred data values, summarized roll-ups, totalized data, and key performance metrics via EIDI system online analytic calculations.

The EIDI Asset Framework (AF) is a database model that hierarchically represents the customer's physical assets in a metadata layer. The AF database includes the sensor data points used to comprise an asset along with run-time asset analytics for derived/calculated data points and unit attributes and provides real-time notifications when unexpected conditions occur.

EIDI system users can also classify specific run times as event frames, which time-slice the data into meaningful periods of time (e.g., batches, lots, startups, shifts). These event frames further contextualize the data for analysis against similar event types.

As such, the raw sensor data input to the EIDI system goes through several iterations and transforms basic sensor data to highly contextualized asset and event relative information, suitable for input to artificial intelligence (AI) and predictive analytics. A user selects which data or event to extract. This contextualized data is prepared and converted into a published data set, ready for input into data lakes, big data analytics, advanced data visualization systems, and predictive analytics, such as ML.

An alternate method of providing data to big data and predictive analytic systems is to stream real-time and relational data directly into a data lake for subsequent use by analytics software. This approach collects and stores data, often in its natural format, into a data warehouse or cloud-based storage, known as a *data lake*. Subsequently, data scientists, IT personnel, engineers, business people, or consultants clean and prepare the data so that it can be input to their analytics. Which approach to use depends on the following criteria:

- What groups are managing the contextualizing and curating the data, that is, are they OT or IT/data scientists? If operations, production, and engineering personnel (OT) are extracting, verifying, and

cleansing the data, it seems to make sense to use an EIDI system in production environments where there can be tremendous payback in terms of decreasing production costs, maintaining an asset fleet, situational real-time awareness, quality improvement, and unified reporting in the production environment.
- If the data is simply collected for analytic purposes, the data lake may suffice. However, because the data is raw and uncontextualized, this requires personnel with high levels of specific domain knowledge to
  - Determine which data is relevant for analytic usage,
  - Clean and confirm the data is accurate,
  - Put the data into its proper context, and
  - Format the data so that it is easily consumable by the analytic systems.

In the long run, this approach may be costly and relies on personnel being extremely familiar with the data, as industrial Internet of things (IIoT) sensor data is not validated in the same way that a traditional control system (distributed control system [DCS], programmable logic controller [PLC], supervisory control and data acquisition [SCADA]), together with a PI System, would. As such, this approach requires substantially more efforts in the following areas:

- Determining what the data actually represents
- Validating the data is accurate; if not, painstakingly cleansing the data
- Determining which data should be discarded as outlying data, meaning data that does not accurately represent what you are trying to analyze or model (e.g., idle equipment, data collected when making out-of-spec product, or transitioning between products)
- Filling in missing data gaps
- Formatting and publishing large cleansed data sets so that the analytics system can easily ingest the data set

We have presented two approaches to corporate or cloud-based analytics. However, it's not an either/or decision. Both can be used in the following manner—the EIDI AF system provides a workbench of sorts for preparing, accessing, evaluating, and putting advanced analytics models into operation. Companies can integrate this AF-contextualized EIDI data into their data lakes for more broad analytics and reporting. This approach also prevents data scientists from having to cleanse, prepare, and remove a large amount of data in a Python environment.

Using a Standard Method for Advanced Analytics Integration

Peter explains that the team chose to use an EIDI-provided integration tool because it *eliminates the following delays,* which are often encountered in advanced analytics projects:

- Developing programmatic extraction scripts, which are often unreliable or developed in a nonstandard way
- Having to contextualize the data in terms of plant, process, product, or specific event
- Producing data sets that contain inconsistent or nonstandard formatting, which requires substantial rework time
- Using inefficient extraction techniques, requiring restarts, excess overhead, or too much processing time
- Having a lack of understanding in what the scripts actually create
- Being overconfident that the data is accurate
- Constantly maintaining or developing new scripts, costing time

Peter noted, "Our operations people and engineers may be skeptical of insights coming from a corporate team that does not have the expertise to make day-to-day decisions on the industrial floor. For now, we need something to accelerate and automate operational data extraction. People will learn how to extract data collected by the EIDI, and they will be prepared to use it with predictive analytics tools, business intelligence tools, and AI tools. The EIDI has become critical to our operations. The scope is becoming larger than we thought, so we need to make sure that extraction tools are standard, easy-to-use, robust, and accurate. These tools must be able to easily extract our data for use elsewhere in the company."

Using these methods, data scientists and analytics teams will have the context that they need to fully understand the operational data. Peter reminds the team that they have emphasized the use of process flow diagrams to create digital plant models (see Chapter 4). This is a vital step in their digital transformation: to have a simple nomenclature for the unit attributes that will be incorporated into their models.

Integrating Production Event-Based Data to Advanced Analytics

Next, Peter has the team review how to extract the data using the production event-framed data for each of the operational states among all the units: "In this case, we can extract any of the time-derived values for each state.

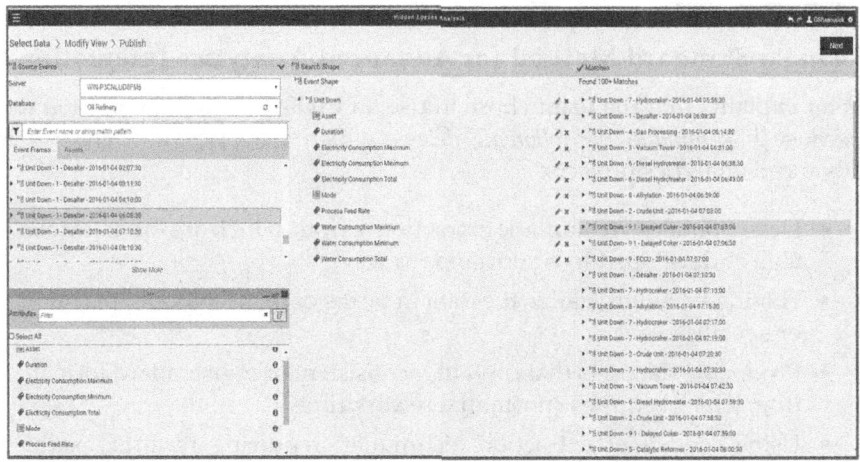

FIGURE 10.4
Event-framed time-series data extracted using the digital plant template.

This derived aggregated information can be reused in real time by AI algorithms using the process mode for enhanced decision-making."

Figure 10.4 shows an entire event-framed data set for all process units in the refinery, aggregating production, energy, water by unit, and by unit operational mode. This data set may be reused by other business intelligence analytic tools.

This standard, configurable extraction tool integrates operational data stored in the EIDI for analysis on Amazon Web Services (AWS) or a Microsoft Azure environment. Supporting data can be retrieved from Hadoop, common relational databases, data lakes/data warehouses, or can be ingested by Amazon Kinesis, Microsoft Azure, or Apache Kafka, and messaging hubs, which are often seen in big data environments.

Peter Argus and Monica Armstrong, who coordinates planning and economics, review their operational industrial workflow, which integrates the work of operations, planning, engineering, and management (Figure 10.5). They agree that the manufacturing process flow and avoiding process constraints are key targets when integrating business and operational systems.

Peter shows the team how to extract the EIDI data using the digital plant template, resulting in the data set shown in Figure 10.6. After successful extraction, Peter feeds the data set into Microsoft Power BI, which creates a cloud-based report. Figure 10.7 shows the entire refinery unit production and consumable losses by unit, operational mode, and by operational shift for an area of the refinery.

In the Power BI report template, several folders were added that describe aggregated KPI information, calculated by the EIDI's real-time analytics using the unit template. The report can display monthly energy consumption

Operational and Business Analytics Integration 237

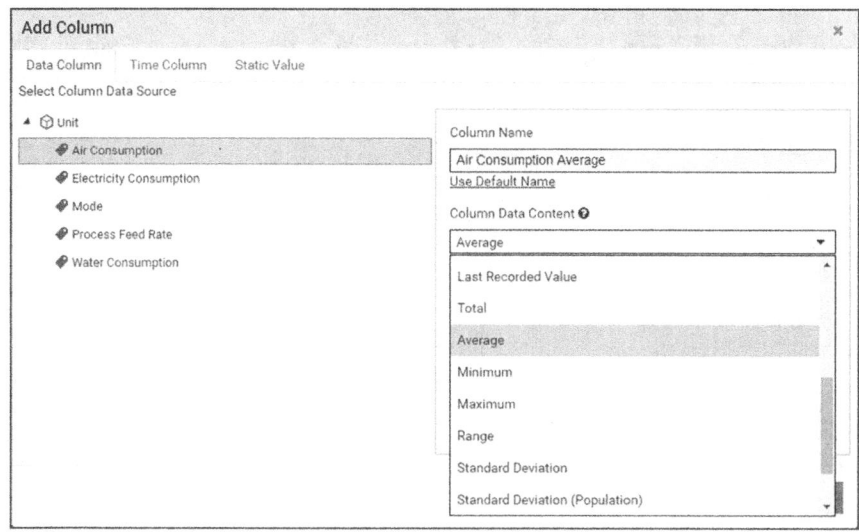

FIGURE 10.5
Adding derived values from operational data.

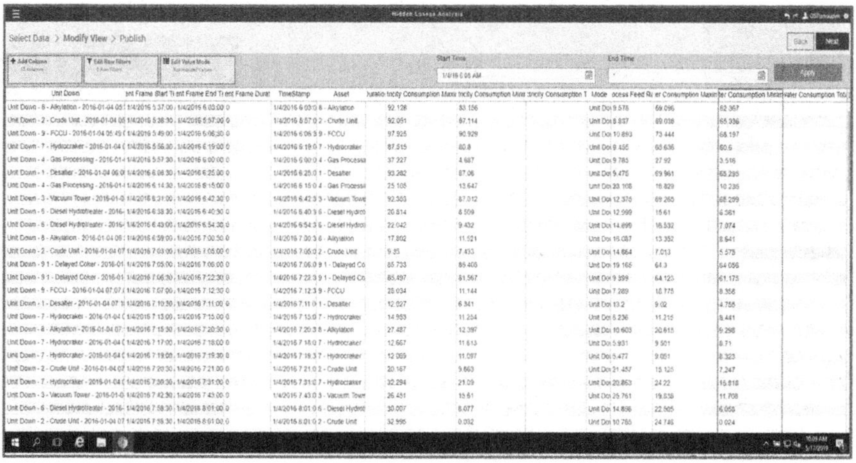

FIGURE 10.6
Refinery digital plant template results extracted for advanced data analytics.

by production area, as shown in Figure 10.8, operating modes, historical production trends, and historical consumption trends (Bascur 2019). As shown here, we can display results and filter by season of the year.

Using operations and production data together allows for better corporate reporting. Peter explained, "We need to ensure that orders are produced on time and in full while optimizing profitability, so that we improve our

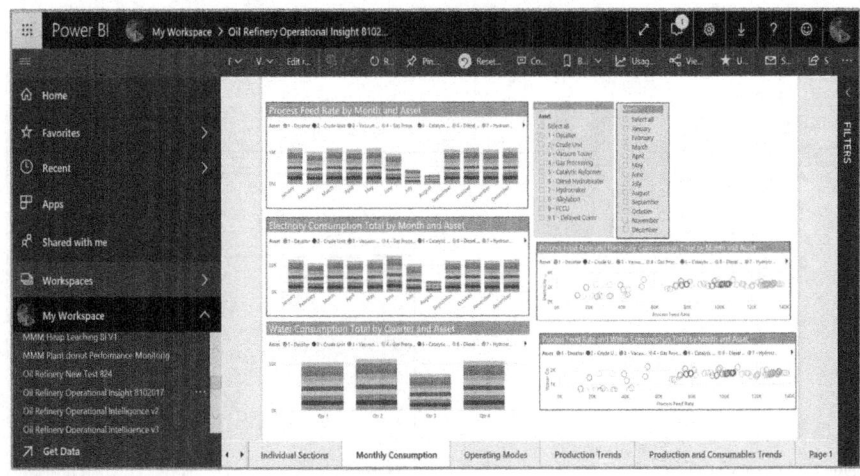

FIGURE 10.7
Microsoft Power BI dashboard folders to present yearly results.

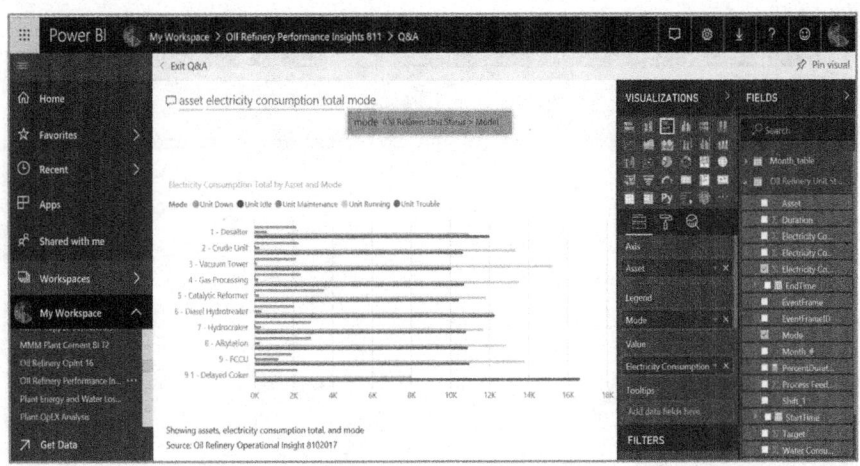

FIGURE 10.8
Microsoft Power BI consumption analysis.

capacity and sequence planning. We can also improve refinery linear programming (LP), also known as *linear optimization modeling*, to get an accurate and timely projection of best yields and consumables."

Monica added: "These AI tools will assist us in analyzing the operations data in order to optimize refinery operations. In addition, we can model customer demand to improve production target settings according to our process capability. We can also do the following:

- Model and manage our consumable resources (energy, water, raw materials).
- Increase and optimize our production schedule through our online equipment health monitoring system.
- Calculate and show material shortages, delays, quality issues, unscheduled asset downtimes, and operating cost variances in real time, so that operations people can make decisions quickly and avoid unnecessary costs and delays."

Creating a Digital Twin for Process Simulation and Modeling

To better diagnose the performance of a refinery, a specific refinery unit, a process plant, or a power-generating station, many companies have replicated the physical process by creating a software representation of the process at an offsite location, usually in the cloud, but it may reside in the corporate headquarters or at a regional office. The purpose of a digital twin is to have a mirrored copy of the process in a safe environment, where models and analytics can query the data and new online calculations can be developed and tested against this digital twin without affecting the actual plant process. By running digital twin data through different models, statistical programs, ML, and other analytics, technical people can detect if the plant is running in line with expected outputs, using engineering principled models, statistical data-based models, or other algorithms used to predict future behavior.

One way to create a digital twin is to simply transfer all sensor data collected via control systems, together with the calculated (derived) variables, so that a near real-time version is available for analysis. Additionally, the data can be updated in real time by transmitting real-time data from the EIDI to the digital twin, so that the digital twin mirrors values in the plant in near real time. Engineering models, ML algorithms, and other predictive analytics are run in this offline, safe environment. Any insights or predictions made by modeling software or predictive analytics can be transmitted back to the actual plant EIDI, so that operations people can visualize how the actual plant process is tracking relative to model predictions that are stored in the EIDI as future data.

The team is now looking at deploying soft sensors to augment data collected from their control systems to enhance ML, digital twins, and data analytics efforts, which they hope will

- Improve their process control performance;
- Evaluate and resolve process bottlenecks;
- Use predictive analytics to more accurately forecast equipment failures, for example, adding sensors to their critical rotating equipment enables more robust, accurate calculations; and
- Generate more useful real-time alerts using the PF-to-F curve strategy referenced in Chapter 6, for pumps and feeders that encounter issues, such as increasing vibration.

Integrating Time-Series Data with Geospatial Systems

In addition to using time-series monitoring of critical infrastructure assets, it has recently become possible to monitor the performance of wide-area critical assets and facilities in real time by integrating real-time operations data into geospatial analytics and displays. Esri, the worldwide leader in geographic information systems, developed a geospatial platform called ArcGIS, which provides a sophisticated, enterprise-scale geospatial information system (GIS) platform, which allows users to create very intelligent maps containing independent layers of data. The data may be static or real-time data, depending on context and the need to update continuously.

Although this may not be critical in a refinery, this marriage of real-time and spatial technologies provides significant benefits in industries where assets are widely dispersed over a large geographic area. Some examples are

- Pipelines,
- The electricity grid and its associated substations,
- Upstream oil and gas fields,
- Water networks,
- Wind turbine farms, and
- Campus environments to monitor energy consumption by individual buildings.

The benefits also extend to monitoring fleets of mobile assets, such as mining trucks, where companies can get real-time visibility into where each truck is currently located and the health of the truck (via time-series data), and use spatial analytics to calculate optimal routes based on dynamic conditions.

Some real-time data infrastructures, such as OSIsoft's PI System, have developed standard off-the-shelf integration methods that help operating companies manage the integration of time-series and geospatial data.

Customers can visualize real-time energy consumption and related cost data by using intelligent maps and dashboards to see the status and condition of building environments on a regional, national, or global basis. By integrating real-time plant or facilities data in geospatial displays and analytics, enterprises can significantly increase their energy usage awareness, recognize data patterns and trends in real time, and utilize powerful spatial analysis tools to explain why issues are occurring.

The EIDI (PI System) AF asset data model, discussed in Chapter 3, can be exposed to the GIS for asset-relative, intelligent map visualization. In addition, asset data from the EIDI system can be integrated with the GIS to utilize spatial (non-time based) analytics. Figure 10.9 represents an Esri ArcGIS intelligent map with KPIs, real-time PI System data, historical trending, and spatial analytics (Lopez 2015).

For more information on integrating real-time data infrastructure with a GIS (Esri ArcGIS), see www.osisoft.com/GIsintegrations.

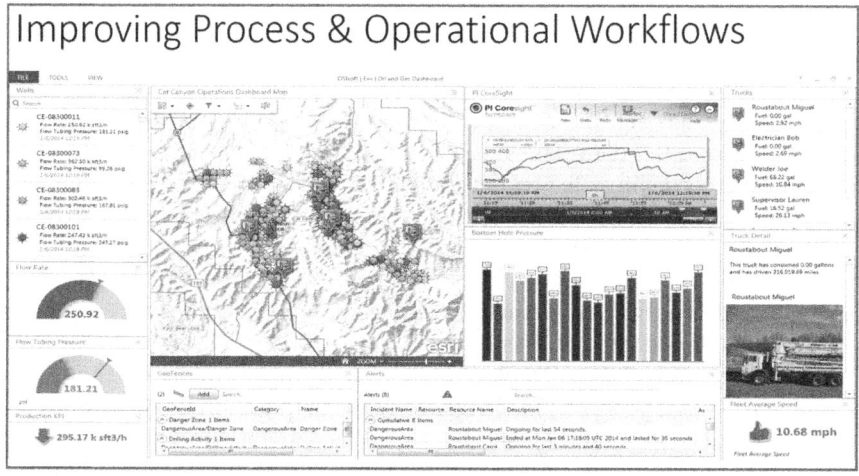

FIGURE 10.9
PI Asset Framework integration with Esri ArcGIS.

What You Should Take Away

The use of time-series data as a component of big data analytics continues to expand. This chapter has presented some of the ways that real-time data collected from the refinery can be leveraged with corporate systems and cloud-based analytic software, such as advanced analytics and visualization software (e.g., Microsoft's Power BI software), ML, programming languages like R and Python, and integration with geospatial systems (Esri's ArcGIS system). These tools provide ways to aggregate time-series operational data with corporate business databases, providing enhanced ways to query information that is generally unavailable at the plant level.

The use of ML to model and predict refinery/plant process and equipment behavior has dramatically increased. Companies use ML to predict critical equipment behavior and estimate failure times. Using this information, plants can more effectively utilize capital for maintenance outlays and better estimate when assets need to be replaced.

The South Texas refinery EIDI data is realizing increased value from these corporate business and advanced analytics by using their highly granular time-series refining data, rather than summarized or averaged data. Consequently, analytics results are much more accurate.

The refinery EIDI system contextualizes data so that it is more effective when using offline analytics and ML. Much of this is because it links data streams to physical assets. It also allows engineers and data scientists to compare and analyze similar event-framed data (i.e., when making identical products or when a specific unit is running in a specific state). This reduces the need to manually inspect and remove outlying data.

The ProcIndustries team developed standard extraction methods that prepare and transmit contextualized refinery data to the analysis software systems. This reduces the need for people to develop their own extraction tools. As a result, ProcIndustries has significantly reduced the time spent extracting, preparing, and formatting EIDI data. This ensures that published data sets are produced correctly and can be ingested by the analytic of choice, reducing time to insight.

Analytics can also model and predict energy consumption under a given set of plant operating conditions. This can help management decide how best to reduce consumption, either by continuous improvements in plant operations, by energy infrastructure replacement, or both.

The South Texas refinery has created and utilizes an offline digital twin. This a virtual copy of the EIDI plant system, with sufficient data to model and simulate the plant process. A mirror image of the plant EIDI is copied to create the digital twin. The EIDI sends near real-time data updates to keep the digital twin current with unit operations. By using a digital twin, engineers can develop analytics or run modeling software that simulates the process and makes predictions. Plant operations personnel can subsequently observe the variance between the predicted values and actual plant performance.

By creating the digital twin, the team can also model refinery processes and yields with ML, highlighting which plant variables are most critical to effective refinery control. Once a model has been developed, operations people can monitor variances in real time, viewing what the model has predicted versus what is actually occurring.

References

Bascur, O.A. 2019. "Process control and operational intelligence in mineral and metallurgical processing." In SME *Mineral Processing & Extractive Metallurgy Handbook*, vol. 1, eds. R.C. Dunne, S.K. Kawatra, and C.A. Young. Englewood, CO: Society for Mining, Metallurgy & Exploration.

Lopez, D. 2015. "PI System and Esri integration overview." Presented at OSIsoft PI Users Conference, San Francisco, CA.

Additional Reading

Afshari, M., and Wang, J. 2019. "PI system for critical operations and advanced analytics." Presented at PI World 2019, San Francisco, CA.

Ammarell, E., and Wang, J. 2019. "Insights: Time-series data science enablement: Pi Integrators and OSIsoft cloud services." Presented at PI World 2019, San Francisco, CA.

Bascur, O.A., and Soudek, A. 2019. "Grinding and flotation optimization using operational intelligence." *Mining, Metallurgy & Exploration* 36(1):139–149.

Ericson, G., Martens, J., and Rohm, W.A. 2018. "What is Azure Machine Learning Studio?" https://docs.microsoft.com/en-us/azure/machine-learning/studio/what-is-ml-studio. Accessed July 2019.

Goldratt, E.M. 2014. *The Goal: A Process of Ongoing Improvement*, 4th rev. ed. Great Barrington, MA: North River Press.

Gopal Krishnan, G., and Hertler, C. 2018. "Fit for purpose: Layers of analytics using the PI System—AF, MATLAB, Machine Learning." Presented at the PI World 2018 Lab, OSIsoft, San Leandro, CA. https://pisquare.osisoft.com/people/gopal/blog/2018/07/11/fit-for-purpose-layers-of-analytics-using-the-pi-system-af-matlab-machine-learning.

Kelleher, J.D., Namee, B.M., and D'Arcy, A. 2015. *Fundamentals of Machine Learning for Predictive Analytics*. Cambridge, MA: MIT Press.

Raschka, S. 2016. *Python Machine Learning*. Birmingham, UK: Packt.

11
ProcIndustries Enterprise-Wide Rollout

I have never met a CEO who does not know the value of all data from his or her assets.

J. Patrick Kennedy

Chapter Overview

This chapter describes how ProcIndustries presents and justifies the business case to their management for scaling the enterprise industrial data infrastructure (EIDI) for the South Texas refinery to ProcIndustries' entire fleet of refineries. The digital transformation team presents the refinery's quantifiable and non-quantifiable returns they have realized since they initially implemented the EIDI. Once management gives approval, we discuss best practices for deploying an EIDI at multiple sites as part of an enterprise-wide initiative.

ProcIndustries Continues Their Digital Transformation

ProcIndustries' operations leadership has recognized the significant cost savings and technological improvements that have occurred at the South Texas refinery. Since the initial enterprise industrial data infrastructure (EIDI) deployment, the refinery achieved expected payback in just over 18 months from the EIDI installation date.

The hard work and success of the past year helped the refinery transformation team to significantly develop and hone their skills, relative to the start of their journey. Peter Argus, the continuous-improvement manager, has built a multidisciplinary team, comprised of process engineers, process control engineers, maintenance personnel, business planning and scheduling people, information technology (IT) personnel, plant management, and others.

Because of their South Texas refinery's success, the ProcIndustries leadership team has given the team a go-ahead to create a proposal supporting enterprise EIDI company-wide deployment. Management expects the team to

- Summarize the South Texas refinery operational improvements, cost reductions, and how the team went about uncovering hidden losses to improve overall refinery effectiveness;
- Explain why a company-wide deployment is required and project the expected benefits; and
- Discuss the plan for enterprise-wide deployment.

With the leadership and support of Bill Roberts (vice president of operations for ProcIndustries), who has seen significant enough variance reductions between production targets and actual refinery output, the team works on the proposal and justification.

Monica Armstrong (planning and economics coordinator) proposes a top-down deployment based on the deployment structure of what they had accomplished at the South Texas refinery, using their standard templated methodology. Bill supports her approach, as he believes people should examine different ways to get people engaged and empowered to try new things. This is the backbone of his digital transformation strategy, and requires changing traditional ways of thinking and planning. Bill wants to use the enterprise-wide deployment to help employees adapt to his continuous-improvement and innovation goals. One objective is to automate repetitive human tasks so that the refineries perform optimally. The second strategy is to challenge people to look beyond current practices and workflows, seeking out new and creative ways of using the data to uncover business value.

The team begins the process of planning for this extensive deployment. They begin identifying the business processes that require modifications and how to foster collaboration at all levels of the organization. Their people/technology/process message centers on the benefits of EIDI technology. The team needs to express their ideas to management succinctly and accurately if they have any chance of company-wide success.

One topic they will present to management is how the integration of EIDI data with other ProcIndustries software systems led to greater productivity and cost savings. Specifically this was done by

- Modeling refinery processes using a digital twin concept, which replicates their physical operations data and uses various process and statistical modeling software to compare actual results against design expectations and to predict future behavior;
- Implementing a proactive condition-based maintenance program that continuously monitors their critical equipment, versus relying

on vendor calendar-based service times. Online EIDI calculations alert plant personnel when an asset reaches a maximum prescribed run time or when an asset's performance is degrading or operating outside of expected parameters. The asset is then subject to a maintenance review or is taken out of service. The EIDI sends notifications to the ProcIndustries' asset maintenance system indicating changes in equipment status;

- Connecting utility meters to monitor and correlate energy consumption to process events and activity. After careful analysis, they have reduced energy usage by 3%; and
- Extracting contextualized production data and publishing as data sets, used by corporate analytics software for yield and run-time cost analysis.

The digital transformation team will present four fundamental steps that ensure the best likelihood of success for enterprise-wide deployment:

1. Build the EIDI business case and obtain management approval.
2. Deploy the EIDI and initial use cases that capture quick returns.
3. Assess the successes of initial use cases. Take corrective action as needed.
4. Administer ongoing governance and prioritize future use cases.

Building the Business Case

The first step begins with the following six objectives:

1. Collaborate with local site personnel to identify key enterprise objectives as a team.
2. Provide economic justification for the fleet rollout.
3. Pinpoint initial use cases that have a rapid expected payback.
4. Achieve buy-in from the management team and key staff-level influencers/team leaders.
5. Foster cooperation and collaboration from all involved. This includes working as a cohesive team, eliminating siloed business practices, and providing sufficient training for people to succeed.
6. Create a project roadmap that identifies and schedules the initial refineries and the initial EIDI use cases. Ensure that the roadmap is periodically updated, based on changing business strategies and the results of completed initiatives.

Financially Justifying the Proposed EIDI Enterprise Rollout

Peter Argus, ProcIndustries' continuous-improvement manager, knew that the next challenge would be one of the most difficult of his career. Successfully implementing an EIDI project at the refinery had been a remarkable achievement. However, taking this to the rest of the enterprise would be an order of magnitude more difficult (Thornton 2019).

Before the EIDI deployment, the ProcIndustries South Texas refinery had been way behind the technology curve and lagged behind their competitors. It took a sizable plant incident for management to understand they needed to enact significant changes to refinery operations. Their first priority was to make sure an event like this never happened again by transitioning to refining-industry best practices. Smart planning, management support, effective deployment execution, and successful teamwork helped the refinery reach their goals in less than a year.

Peter recalled that when his former company opted for an enterprise-wide transformation, many of the departments did not work together at all. Senior managers vigorously defended their turf and were resistant to collaboratively working together.

Corporate business management, marketing, corporate IT, and the refinery managers were highly disconnected. Many people were protective of their data and reports. If information was shared at all, it was sent via spreadsheets or through after-the-fact reports. There was no comprehensive strategy on how to centralize operations and production data and how to make the data easily available to knowledge workers when they needed it. Peter contemplated whether he would fight the same battle at ProcIndustries.

The digital transformation team reconvened to discuss how they would justify enterprise-wide deployment of the EIDI across all refineries. They determined that they would split the proposed benefits into two categories: estimated quantifiable returns, such as energy and maintenance savings; and those not easily quantifiable, but equally as important in achieving corporate objectives and minimizing risk to the enterprise.

The team thought the best method to calculate potential returns and benefits was to interview a cross section of employees from each potential deployment site as well as several key corporate employees to formulate a comprehensive view of how the EIDI could be of most value to everyone. This way, change is initiated in a productive way.

The team asked each of the interviewees the following questions:

1. What were their top pain points? Which problems occurred most often?
2. How would access to relevant data assist them?
3. How would this data help the refinery solve its problems?

4. Which issues should be initially addressed during EIDI planning and deployment?
5. What are the possible benefits and returns from having the data readily available?

The interviewees for each site consisted of the following personnel:

- Vice President of Operations for ProcIndustries Bill Roberts
- South Texas Refinery Manager Tom Jordan
- Production Manager Tim Olsen
- Planning and Economics Coordinator Monica Armstrong
- Operations Manager and Shift Supervisor Jesus Gonzalez
- Maintenance Manager Paul Morgan
- Process Control Manager Chuck Smith
- Process Engineering Manager and Process Engineer Alex Moretti
- Energy and Utilities Manager Ron Erickson
- Process Safety and Environmental Manager Raj Singh
- Product Quality and Laboratory Manager Chen Wang
- IT Manager Paul Verlaine

After completing those interviews at the various sites, the team interviewed several corporate personnel who would utilize the aggregated site data or have a key role in the fleet-wide deployment. These are the directors of

- Fleet-wide business operations;
- Fleet-wide production planning and scheduling;
- Refinery operations;
- Corporate IT;
- Fleet-wide maintenance;
- Corporate engineering; and
- Health, safety, and environmental engineering.

After four weeks of comprehensive interviews, the team reconvened to summarize their findings and list the potential benefits and returns for the rest of the fleet. Peter instructed the team to combine the interview conclusions with the results achieved by using the EIDI at the South Texas refinery. Some of the projected benefits were compelling, but the team could not quantify them into a specific return-on-investment (ROI) number. As a result, they divided their proposed justification into quantifiable and non-quantifiable items.

Quantifiable Benefits
1. **Process improvements.** By having real-time situational awareness, operators and engineers are able to resolve issues before they become large, expensive problems. These include
    - Detecting issues that increase variance in actual production rates versus planned targets; then correcting those issues, some of which included correlation analysis, process improvements, and optimizing distributed control system (DCS) control loops. This resulted in just over a 0.5% gain in barrels of oil produced.
2. **Increased operating margins through reductions in refinery operating-related expenses** by transitioning from a calendar-based to a condition-based or run-time-based maintenance program in which the EIDI monitors real-time equipment performance.
   Averting replacement of new equipment because of improved maintenance practices frees up capital for other refinery investments. The team estimates the following savings:
    - Increased equipment availability, with an expected uptime of more than 98%. Online calculations will immediately identify assets that are not performing adequately, require checking, need repairs, or are taken out of service. Based on experiences at the South Texas refinery, the increased equipment availability reduced operating costs by 6% annually.
    - Total lost production resulting from lack of equipment availability is estimated at $35 million annually, based on previous experiences at the South Texas refinery.
    - By avoiding unscheduled equipment failures, maintenance expenses and associated labor costs are estimated to decrease at each facility by 4% annually, based on South Texas refinery savings.
    - By having the EIDI identify the assets most prone to failure, budgeting will become much more data driven, resulting in more accurate methods of deploying capital to replace or repair critical assets. The expected savings is 6% per year in each refinery/plant's maintenance budget.
    - Major outages will be able to be flexibly scheduled, based on market conditions, rather than by rigid scheduling or equipment failures. The South Texas refinery found this resulted in 1% annual revenue growth.

3. **Reduced operational expenditures (OPEX) through energy savings.** Having real-time visibility to energy consumption will enable the following:
   - Track actual energy costs for each barrel of energy produced.
   - Identify which assets are consuming the most energy or excessive energy and correcting the underlying problems.
   - Reduce steam traps caused by broken pipes, tubes, vessels, and so forth.
   - Reduce compressor energy usage caused by equipment fouling, water contamination, and so on.
   - The expected reduction in energy costs per refinery/plant is 4% per year.
   - In addition, the validation of fuel and utility custody transfers (gas, power) is estimated to save 1% per year.
4. **Increased margins through supply chain savings.** Directly integrating real-time production and operations data into the corporate software system will have the following expected outcomes:
   - Tracking raw material and energy consumption in that system, making it available for accounting personnel
   - Transmitting equipment status to the corporate maintenance system so that the work-order process is much more efficient
5. **Reduced OPEX through improved management, technical staff productivity, and reduced travel costs.** Division managers, vice presidents, knowledge workers, and others will have real-time visibility into their fleet operations from dashboards accessible from desktop or mobile devices. Knowledge workers will reduce their time spent collecting data from disparate siloed systems by 20% to 25%. In addition, engineers, research and design (R&D) personnel, and others would not need to travel to the sites to diagnose and fix problems. The estimated travel cost savings per facility is $100,000 annually.
6. **OPEX savings through reduced cost of EIDI deployment.** The South Texas refinery created an asset database with hierarchical object templates. These asset templates contain online calculation and notification alert capability. By reusing these templates, much of the EIDI configuration time for each site can be reduced by 40%.
7. **Reduce capital expenditure (CAPEX) and OPEX through decreased software costs for new use cases.** Some new analysis use cases may occur where it is more feasible to place similar data into

a separate EIDI system, such as monitoring condenser performance for the entire fleet. With the EIDI asset template capability, new asset databases can be generated very quickly. In addition, actual plant data is replicated to a new EIDI in near real time. Using a separate EIDI is beneficial in cases where software development or significant data retrieval queries are needed, so as not to compromise the performance of actual plant control networks.

Non-Quantifiable Benefits

1. Monitoring of operations and production activity for the entire fleet using a common real-time data infrastructure in which the EIDIs would act as systems of record in all refineries. The EIDI would monitor the entire value chain using the same technology and data access. Future use cases could leverage existing EIDI data once deployed on an enterprise-wide basis. This would significantly reduce project costs versus specifying a real-time data platform for every site.
2. Providing a safer and environmentally sound workplace by continuously monitoring refinery/plant conditions and emissions. The company will use the EIDI to track sustainability initiative progress. For the plants, it can calculate actual emissions versus allowable emissions for a given time period and alert personnel if exceeding maximum allowable thresholds.
3. Standardizing the way employees access and analyze data as well as how reports are generated for both internal consumption and external compliance reporting. This would increase productivity by reducing unnecessary time to acquire and analyze data from different systems.
4. EIDI standardization reduces the need for corporate IT to deploy and support multiple EIDI platforms. Software systems can ingest and transfer data through reusable, non-customized methods, requiring fewer corporate IT resources to develop custom interfaces. Training costs also decrease when utilizing and supporting a single EIDI platform.
5. Integrating plant and refinery information with corporate business systems, such as SAP and big-data analytics such as Tableau and Microsoft Power BI. These tools would become more effective by utilizing highly contextualized operations, production, and equipment performance data to achieve strategic corporate objectives. The data could also be placed into a data lake for other analytic software systems.
6. Providing secure access to EIDI information via mobile devices to appropriate personnel to improve productivity and awareness when away from their office or workstation.

7. With each EIDI acting as the refinery or chemical plant system of record, the long-term highly granular historical data archive would provide accurate information for possible inquiries such as emissions activity, regulatory audits, cybersecurity audits, and information requests for litigation. Enabling best practices would avoid public scrutiny.
8. The EIDI systems would benchmark existing refinery or chemical plant performance and would continuously monitor improvements made as well as subtle long-term degradation of assets.
9. The ProcIndustries upper management team would improve collaboration among departments by utilizing the EIDI platform for their own data analysis as well as for interdepartmental initiatives. Siloed data barriers would be eliminated, as the EIDI would become the central real-time data repository.
10. Information collected by the EIDI would create an information archive on how best to operate the facilities. This would benefit ProcIndustries when plant operators, engineers, or maintenance employees leave or retire. The refineries or plants could more quickly train new employees.
11. Using the EIDI's asset framework (AF) template object models would significantly reduce the time to roll out an EIDI at each facility, leveraging and duplicating what was implemented at the South Texas refinery. Newly acquired or newly built facilities could quickly deploy the EIDI.
12. It would be easier to recruit new technical people, especially recent graduates by enabling them to use the latest technology to examine data and create modern-day analytics.

Management's Decision

The team presented their findings to the ProcIndustries' leadership team. They presented current operations methods; identified key problems that were obstacles to operating more effectively and efficiently; and discussed costs versus expected benefits, how the rollout project would be implemented, and their plan to get everyone's buy-in through ongoing communications, sufficient training, and increased collaboration.

As with Peter's former employer, there was skepticism regarding several items:

- Is the total cost of the enterprise-wide deployment of the EIDI worth it?
- Are the ROI calculations close to being accurate?
- What metrics ensure success of the initiative?

- Who will take ultimate responsibility for the project?
- What will the role of the infrastructure provider be?
- How much labor will be required by ProcIndustries personnel?
- Where are the most significant risks?

As the meeting continued, the team was not sure whether management would approve it or not. Some were skeptical of the validity of the ROI numbers, some wanted to use capital funding for newer equipment, others were not convinced that the departments and sites would work together to ensure project success. However, the discussion shifted to scrutiny of the ROI numbers. As deliberations continued, leadership grew more comfortable with the ROI targets that had been signed off at lower levels. Peter assured them that his team would own the infrastructure deployment and realization of the proposed ROI, mitigating management's received sense of risk.

In the end, management approved the expenditure for the following seven reasons:

1. The risks associated with not going forward were potentially more severe that the project cost. These risks included another site incident, an environmental catastrophe resulting in regulatory actions and increased public scrutiny, possibly leading to company valuation issues.
2. Each site needed a reliable real-time data system of record, with accurate historical data.
3. Too many improvement initiatives, including big-data and predictive analytics, could not effectively be done without easily accessible, accurate data from the production sites.
4. Energy and maintenance costs for large assets were spiraling upward. The proposed condition-based monitoring of critical assets would help to reduce those outlays.
5. ProcIndustries could potentially lose market share because of poor yields or inferior production practices versus their competitors.
6. Management had constantly been disappointed with political infighting and turf wars. Perhaps the enterprise-wide EIDI would act as a catalyst for collaboration across the sites, and between the production sites and corporate personnel.
7. Management desperately needed more accurate and timelier information from the sites, so they could have easier and quicker visibility into fleet operations.

Management also agreed with the implementation plan that included the use cases described earlier in the chapter as justification for the EIDI enterprise-wide rollout. Figure 11.1 highlights the use cases for enterprise-wide EIDI implementation.

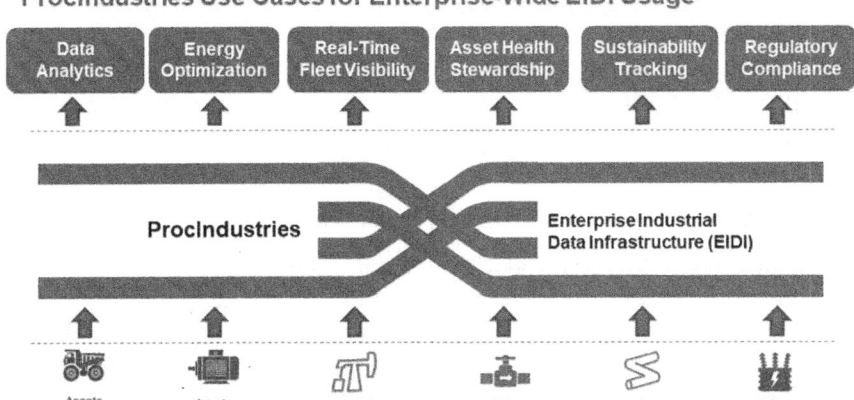

FIGURE 11.1
EIDI enterprise-wide use cases.

Deployment of the Infrastructure and Initial Use Cases

Once ProcIndustries management approved the enterprise-wide EIDI scale-up, they announced that the project team that initially deployed the South Texas refinery EIDI would remain in place to define and implement the roadmap for deployment throughout the entire enterprise. In addition, some corporate personnel from the business operations group, corporate IT, and project planning were added to round out and assist the team.

Shortly thereafter, the EIDI enterprise rollout team convened and identified eight key internal areas to address prior to meeting with OSIsoft, their software data infrastructure provider:

1. Identify dedicated ProcIndustries resources to manage the new EIDI platforms. One IT resource and one engineering/businessperson from each of the refineries will work with the team.
2. Define standard architectures, standard configuration processes with customizable templates.
3. Meet with local process control engineers to discuss EIDI data sources for each refinery.
4. Review South Texas refinery key production indicators (KPIs) with local refinery management teams. Customize as needed.
5. Define EIDI user-training content and how to administer training at each refinery.

6. Build a cross-site support team to develop best practices and to assist each other.
7. Measure adherence to EIDI rollout standards as means of assessing progress and risk.
8. Keep management regularly informed of rollout progress.

Working Toward a Smooth EIDI Enterprise-Wide Rollout

The team met with their software data infrastructure provider, OSIsoft, to define rollout requirements and to map out the deployment process. They discussed their initial EIDI implementation at the South Texas refinery, which use cases they deployed, and benefits and takeaways from that experience. The participants then discussed implementation requirements and scheduling for the next set of refineries.

OSIsoft announced that they would assign a dedicated customer success manager (CSM) to work together with the ProcIndustries team, specifically Peter Argus, to manage the rollout and help achieve ProcIndustries' business objectives and goals. The CSM would manage the EIDI architecture design, implementation, training, and coaching on how to achieve optimum results.

The CSM also leverages subject matter experts to accelerate the deployments, specifically enterprise software architects, PI System installation engineers, software product specialists, and experts in refining EIDI best practices. Figure 11.2 shows the CSM's responsibilities.

Key Activities in the EIDI Rollout to All ProcIndustries Refineries

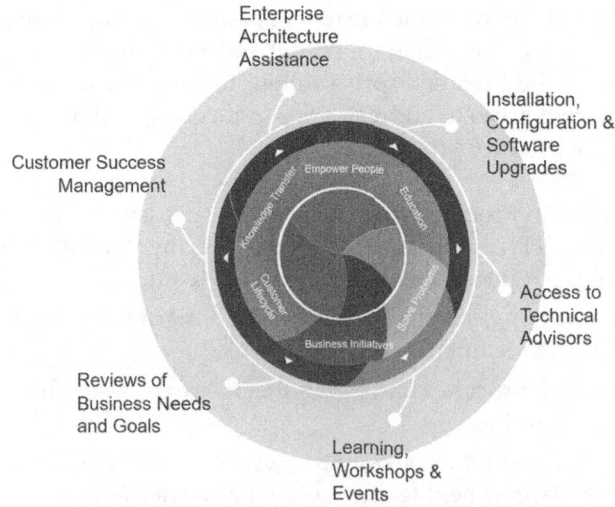

FIGURE 11.2
Responsibilities of the EIDI software vendor in an enterprise-wide rollout.

FIGURE 11.3
The role of the customer success manager.

Figure 11.3 summarizes the resources a CSM brings to the table (Stawiarski and Molano 2018). The CSM's responsibilities are as follows:

- Lead the customer onboarding experience to ensure that ProcIndustries knows all the OSIsoft resources that are included and how to take advantage of them.
- Drive architecture planning, system implementation, and monitoring by developing a strategic multiyear roadmap, architecture plan, and value-realization plan. Together, they plan rollout activities, installations, configuration activities, and recommend continuous-improvement initiatives to increase the EIDI value and returns.
- Identify and document ProcIndustries' KPIs that will allow the progress and status of the rollout, and subsequent data projects, to be measured and reported on.
- Develop long-term strategic roadmaps that are updated annually to meet ProcIndustries' business objectives and address persistent pain points.
- Participate in communication between ProcIndustries and the infrastructure provider to ensure that the final solution meets with OSIsoft best practices.
- Arrange and conduct periodic status meetings.
- Determine subsequent EIDI rollout plans and best practices for periodic upgrades.

- Leverage the OSIsoft technical advisor (TA) team for developing effective architectural plans and provide best practices consulting to implement priority use cases (e.g., condition-based maintenance, event frame production analysis, energy management, extracting data for big-data analytics).
- Organize staff training, focused customer workshops, and other activities, which ensure that ProcIndustries' employees make optimal use of the EIDI.
- Prepare quarterly and yearly assessment reviews and recommend new opportunities for improvement and increased business value.
- Assist in targeted conversations with ProcIndustries to find additional opportunities for the EIDI.

Initial Rollout Activities

The CSM had begun collaborating with the ProcIndustries enterprise-wide rollout team, discussing the best ways to make the initial rollouts successful. For the next few weeks, the team considered the CSM's suggestions. At their next meeting, Peter presented their enterprise rollout plans:

1. Follow the lead of the South Texas refinery, which has updated their process block and process control diagrams based on latest Occupational Safety and Health Administration (OSHA) 29 CFR 1910 compliance regulations.
2. Define KPIs for each site, each unit, and by discipline (e.g., production, maintenance, energy usage). Aggregate and roll up information across the enterprise.
3. Monitor key production metrics for each unit: feed rate, energy, fuel, water, yield, flows, and so forth.
4. Monitor real-time equipment performance as they had done at the South Texas refinery to ensure that pumps, compressors, turbines, condensers, chillers, and other critical assets operate within acceptable ranges and issue alerts when they diverge from these ranges. Ultimately, send these alerts directly to the ProcIndustries asset management and maintenance system.
5. Monitor key raw materials and product quality variables. The team plans to integrate each site's laboratory information management system (LIMS) data into that refinery's EIDI.
6. Establish baselines for the refinery's energy and water consumption. Monitor vessels, tubes, pipes, condensers, and columns for energy leaks, fouling, and contamination in order to reduce unnecessary

energy consumption. Quantify and store incremental improvements. Communicate these successes to management and the sustainability team.
7. Establish connectivity to the ProcIndustries' production scheduling system.
8. Configure the EIDI using the same AF database approach in which the team configures a hierarchical model of the data for each site, reusing much of the asset templates that contain equipment information, input data source, online calculations, and alerting information for each type of asset.
9. Develop and implement the digital plant templates used at the South Texas refinery, tailored for each refinery. Define production event frames that leverage the digital plant template.

The CSM reviews the preceding items and recommends that the teams schedule two jump-start workshops. The first workshop will be conducted to review ProcIndustries' specific business goals, with recommendations on how best to design EIDI connectivity and workflow to accommodate Peter's business objectives. The second jump-start workshop will cover best technical and configuration practices to derive these business values.

The team provides a general timeline for the enterprise-wide rollout for each refinery and the course is set. Figure 11.4 shows a typical workflow for enterprise-wide EIDI deployments.

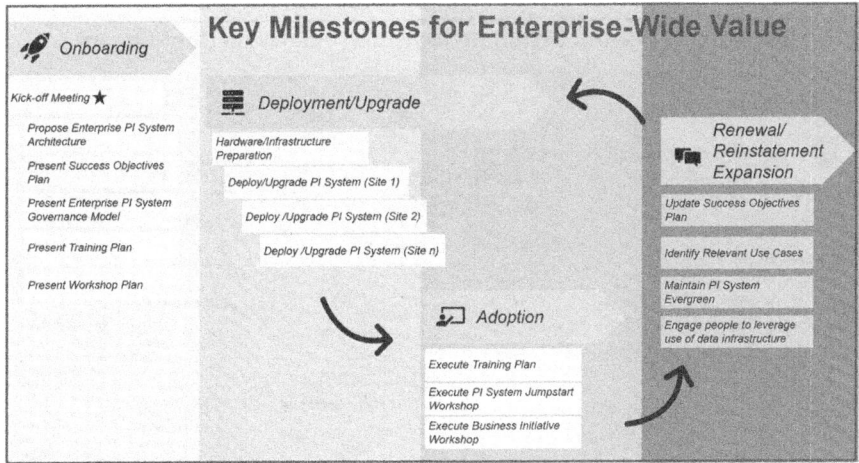

FIGURE 11.4
Life cycle of EIDI rollouts across the enterprise.

EIDI Architecture and Cybersecurity

Once ProcIndustries leadership approved the enterprise-wide EIDI rollout, their IT department started investigating the integration of all process control systems with the laboratory and equipment monitoring systems. IT Manager Pat Verlaine has been concerned about the safety of the networks and about cybersecurity threats, once the operational technology (OT; refinery systems) and IT systems converge on a larger scale.

To meet the needs of security and data reliability, Pat determined the need to deploy redundant EIDI servers for refinery operations and production data using a demilitarized zone (DMZ) strategy. He also saw a need for a mirrored site outside the DMZ. During enterprise-wide rollout meetings, the team stated the need for separate test and development servers, so that EIDI software upgrades and cybersecurity updates are more easily integrated. IT/OT cybersecurity convergence has the potential to solve many industrial cybersecurity challenges (Kanellos et al. 2019).

Phase III: Ongoing Governance and Future Use Cases

To keep an EIDI system in peak condition, it requires proactive, planned software updates and vendor technical support as needed. Figure 11.5 shows the timeline for an EIDI enterprise-wide rollout, displayed as a Gantt chart to

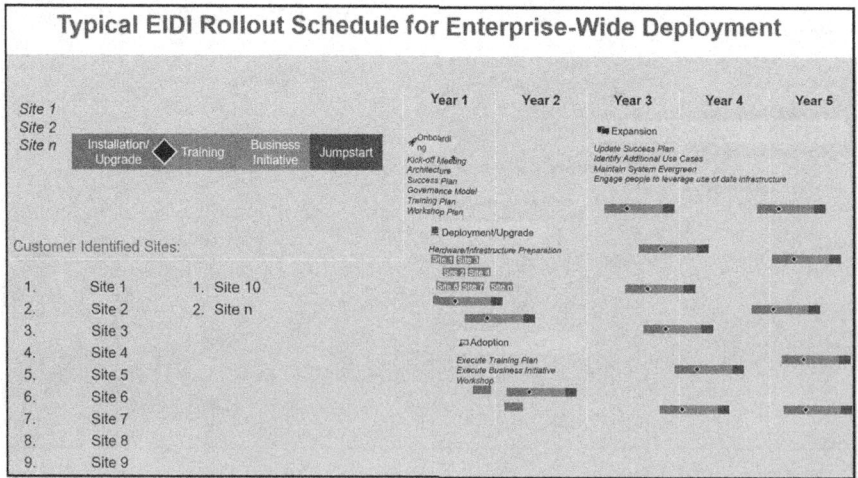

FIGURE 11.5
Gantt chart projecting master rollout schedule.

track implementation and rollout activities. Installation and upgrades, training, business review meetings, and jump-start workshops are coordinated for all stakeholders.

One key to ensuring success on a large-scale initiative such as this is to have quarterly and annual account review meetings to track progress, plan for upcoming major activities, determine if management is pleased with the results so far, document all EIDI benefits and returns, and review any existing technical initiatives or open issues.

Future Use Cases

The beauty of an EIDI is that once it is set up and collecting data, the individual use cases need not be developed all at once. Because the EIDI acts as a system of record at the refinery, it contains a permanent archive, which can be used to populate calculations, analytics, reports, process models, and other use cases to be developed in the future. This allows companies to develop technical or business applications planned for the future by leveraging existing EIDI data. Because deployment teams cannot tackle everything at once, it makes sense to initially implement defined use cases in order to accomplish a known objective. Less obvious use cases, or use cases in which the metrics or analytics are emerging, can be designed and installed later, leveraging existing data. A software infrastructure, versus a defined project, allows companies the flexibility to define and deploy future applications without modifications to the existing infrastructure. Figure 11.6 represents this concept of "value now, value over time."

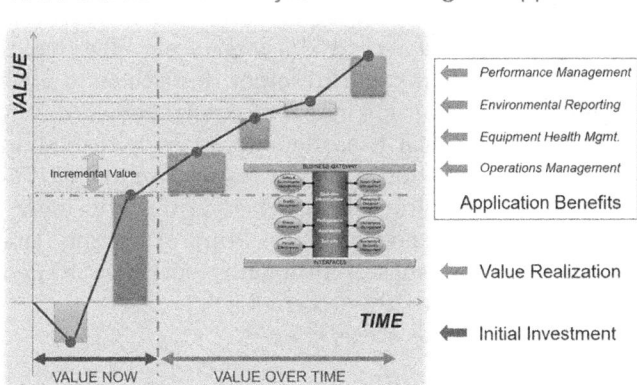

FIGURE 11.6
Extended EIDI value over time versus value limited by a single project approach.

What You Should Take Away

Companies that have deployed an EIDI infrastructure throughout their facilities have realized high ROIs from real-time fleet visibility, improved decision-making, process and quality improvements, energy optimization, and improved enterprise-wide equipment uptime via a condition-based maintenance program.

The EIDI strategy, coupled with continuous-improvement loops methodology, creates an environment for fast, cheap innovative solutions (Kennedy and Bascur 2002). As such, the digital transformation results realized at the South Texas refinery have convinced management to approve an EIDI rollout to all ProcIndustries refineries. Using the EIDI to catch hidden losses, maximize production, and identify root causes will continue to create constant value using the roadmap shown in Figure 11.6. This approach allows companies to survive and better compete in the digital era (Bascur 2019).

Once the concept has been proven and initial pilot sites achieve significant value, many companies elect to deploy an EIDI throughout their fleet. This enables enhanced real-time visibility into all operating facilities. It also provides a common methodology for

- Ingesting data from control systems, industrial Internet of things (IIoT) sensors, supervisory control and data acquisition (SCADA), and so forth;
- Viewing data on desktop and mobile devices;
- Extracting data for advanced analytics and process models;
- Programmatically accessing data for user-developed applications; and
- Interfacing to third-party software solutions.

The value and returns are significant, but often are not easily quantifiable. For management to approve and fund a large-scale EIDI deployment, it is necessary for team members from different disciplines to estimate quantifiable and non-quantifiable returns from enterprise-wide usage. In their funding justification proposal to leadership, team leaders should emphasize the increased collaboration that will result from using the same EIDI software. Knowledge workers will also be significantly more productive simply because they will spend much less time collecting data and more time solving key business issues. Additionally, workers will spend less time arguing which version of the data is correct.

Once funded, the next steps are as follows:

1. Form a multidiscipline, multi-facility team with as many skills as needed for a large-scale initiative.

2. Make sure the vendor appoints a CSM, who will oversee the entire deployment, training, and consultation to ensure success. This person will leverage the appropriate subject matter experts in the vendor company to provide expertise when required.
3. Make sure IT experts are included when designing and architecting a solution of this scale. They will advise how to optimize data traffic and provide best cybersecurity practices.
4. Prioritize which use cases to implement first. Rome wasn't built in a day. Neither will your enterprise-wide deployment. Tackle items that have management's attention and where quick returns are likely.

Once the infrastructure is collecting data, any planned or unplanned use cases to be implemented in the future can leverage their EIDI investment by using existing high-fidelity data. By constantly improving operations through data-driven decision-making, companies can change their culture, becoming innovative and well on their digital transformation journey.

References

Bascur, O.A. 2019. "Process control and operational intelligence in mineral and metallurgical processing." In *SME Mineral Processing & Extractive Metallurgy Handbook*, vol. 1, eds. R.C. Dunne, S.K. Kawatra, and C.A. Young. Englewood, CO: Society for Mining, Metallurgy & Exploration.
Kanellos, M., Owen, B., Ginter, A., Cowell, M., Miller, J., and Barnier, G. 2019. "The growing ecosystem for OT security." Webinar. San Leandro, CA: OSIsoft. www.osisoft.com/about-osisoft/webinars/the-growing-ecosystem-for-ot-security/.
Kennedy, J.P., and Bascur, O.A. 2002. "Influence of computer and information technology on process operations and business processes—A case study." In *Chemical Process Control VI: Assessment and New Directions for Research: Proceedings of the Sixth International Conference on Chemical Process Control*, eds. J.B. Rawlings, B.A. Ogunnaike, and J.W. Eaton. CACHE Series, vol. 98. New York: American Institute of Chemical Engineers (AICHE). pp. 7–11.
Occupational Safety and Health Administration (OSHA). Occupational Safety and Health Standards. 29 CFR 1910. Washington, DC: OSHA.
Stawiarski, A., and Molano, A. 2018. "Enterprise agreement value customer success." Presentation, OSIsoft, San Leandro, CA.
Thornton, C. 2019. "Digital transformation: The power and importance of a digital strategy and implementation road map." Presented at the AISTECH Conference, Pittsburgh, PA, March 26–29. www.aist.org/AIST/aist/AIST/Conferences_Exhibitions/Training_Seminars/DT%20presentations/27-Thornton.pdf.

Additional Reading

Bascur, O.A., and O'Rourke, J. 2019. "Measuring, managing and transforming data for operational insights." In *Smart Manufacturing: Concepts and Methods*, eds. M. Soroush, M. Baldea, and T. Edgar. New York: Elsevier, forthcoming.

Kennedy, J.P. 2018. "The driving force of industry." *Scale and Scope* (September/October): 34–37. www.isa.org/intech/20181005.

Kennedy, J.P. 2019. "Retrospective: Then and now—Looking toward the future." *Hydrocarbon Processing* (January). www.hydrocarbonprocessing.com/magazine/2019/january-2019/columns/retrospective-then-and-now-looking-toward-the-future.

Nazer, Y., and Waray, W. 2018. "Automating continuous process operations." *Intech Magazine* (July/August): 24–29.

OSIsoft. 2018. "The paper mill of the future: Why change agents are the key to digital transformation" (white paper WPPMFEN-092618). San Leandro, CA: OSIsoft. www.osisoft.com/whitepapers/.

OSIsoft. 2019a. "Support for enterprise agreement (EA) program customers." www.osisoft.com/support/value-of-enterprise-wide-deployment/support-for-ea/.

OSIsoft. 2019b. *A New Era in Mill Analytics: How a Data Infrastructure Can Improve Pulp and Paper Production Management and Asset Performance*. San Leandro, CA: OSIsoft. www.osisoft.com/whitepapers/.

12

The Future of the Digital Enterprise

Never make predictions, especially about the future.

Casey Stengel

Chapter Overview

In this chapter, we review the progress the ProcIndustries team has made in their quest for operational excellence and some of the plans they have for initiatives such as cloud strategy, IoT (Internet of things) sensor integration, edge analytics, the refinery and control system of the future, and collaborative manufacturing. We also touch upon some refinery specific applications that become more powerful with contextualized enterprise information data infrastructure (EIDI) data. To begin the chapter, we discuss the traits shared by companies that have best navigated their digital transformation journeys. Finally, we will share the outcome of what happens to the ProcIndustries digital transformation team.

How the Most Successful Companies Achieve Success

We have witnessed hundreds, if not thousands of companies, strive to successfully transform their business using digital technology. Many of them, across all industries, are very successful. Others struggle for years and spend unnecessary resources in quest of their goals. Why the difference? The software they use is similar to their peers. Most entities have sufficient funding and human resources to succeed, but some do not.

In our opinion, there are six common traits of entities (companies) that best harness and utilize digital technology to achieve their business objectives:

1. Align business goals with digital strategies.
2. See operations and production data as critical strategic assets.
3. Team up with strategic partners.

4. Architect for the unexpected.
5. Use analytics effectively.
6. Create self-serve access to needed information.

Alignment of Business Goals with Digital Strategies

These companies have strategic leadership and align every technology initiative with tangible business objectives. Examples of these business drivers are to

- Increase production rates over a given time period;
- Eliminate controllable losses by process optimization;
- Improve product quality or reduce variability over multiple production runs;
- Increase product yield, eliminate waste, scrap, or rework needed;
- Ensure that equipment is always in peak operating condition to optimize maintenance costs;
- Reduce the total cost of production, including energy costs;
- Improve real-time situational awareness and decision-making;
- Make sure production is agile and efficient in order to respond to market changes; and
- Plan for future growth and product demand.

Notice that the preceding items do not include a technology objective. The most effective companies do not deploy technology for the sake of technology. There are always new information technology (IT) trends and technologies that are hyped by thought leadership, consultants, or in IT publications. Sometimes, the technology and the intent are plausible, but the technology is too new, too early for wide-scale deployment, or not easily integrated with existing legacy systems. The underpinning of any technology initiative should be focused on the financials; the product made; worker productivity; or environmental, health, and safety aspects of making product.

The successful enterprises undergo a digital transformation led by the business teams on the operational technology (OT) side. They have the most knowledge of the core business and its challenges. They also are experts in that company's work processes and where improvement is needed. IT departments provide a vital role of identifying and deploying the proper solutions, making sure they are scalable, reliable, and secure. As stated previously, business goals always govern technology initiatives. The core of what these successful enterprises are seeking is operational excellence, generally with a customer-centric focus. They generally do not overcomplicate their journey by attempting too many things at once.

Management's business transformation goals must be culturally aligned with all employees so that everyone is working together to improve work processes and reduce inefficiencies.

Viewing Operations and Production Data as a Critical Strategic Asset

Successful companies view their critical operations data as a strategic and unique asset, ultimately becoming the lifeblood of their operating divisions. Enterprises that have a long view and value their data generally tend to extend their critical data infrastructure to monitor as much of their value chain as possible. For oil and gas, this encompasses upstream exploration and production, midstream pipelines, downstream refining, and into petrochemical manufacturing.

Enabling a critical data infrastructure across this extended domain allows companies to integrate, abstract, normalize, and contextualize information so that it becomes a self-serve model for individuals, software systems, and applications to ingest the information they need.

This critical data infrastructure must be scalable, secure, open, and reliable.

Successful Teaming with Strategic Partners

Enterprises that have great technical leadership know the map of technology providers well. These providers have varying business models. Some suppliers provide a suite of products and applications suited to a particular industry, while others provide a single solution across many vertical industries. Some, as in the case of the EIDI discussed in this book, provide out-of-the-box, standard software without the need for custom programming to start monitoring production and operations activity. Others, such as software models, often rely on consulting and custom services to best provide an effective outcome for the customer.

Our view is that the solution or platform of choice should be purchased from a responsible vendor with a sustained track record in the industry. If they have provided a technology for decades, this technology is their lifeblood. The supplier is unlikely to discontinue an offering the customer has purchased simply because it is not in the best long-term viability of the supplier.

What we have learned over the years is that successful enterprises usually standardize on a technology provider for its critical data infrastructure. To re-engineer this for each asset or on a project-by-project basis is very costly and very human resource intensive. Things that are not easily calculated are total life-cycle costs for having different suppliers, such as integrating data into other critical software systems like maintenance and financial systems.

A relationship between a large manufacturing enterprise and its key software suppliers generally lasts decades, perhaps 30–40 years, or longer. When selecting something as strategic as this, great technology is important. But even more important is mitigating risk by selecting the right business partner. Failure to do so will be far more catastrophic than the encountering gaps in technology over the life of the software.

Architecting for the Unexpected

When deploying an EIDI system across one or more operating facilities, it is essential that subject matter experts, business users, and software applications are able to access and ingest information about various aspects of the operating facility without having to know the unique features of the plant equipment, its control system, or its instrumentation nomenclature. The way to do this is through an asset framework (AF) model that we have described in the various chapters of the book. Through use of these meta-data models, a layer of standardization is created for consistent data access, online analytics, and real-time notifications. The mechanism to accomplish this is to build standard, reusable templates (think of them as Lego blocks) so that portions of an operating facility or an entire plant can be reconstructed virtually as a digital twin in the span of weeks or months, not years.

Over the past few decades, we have seen many enterprises revise or refocus their strategic business objectives. To accomplish these goals, many companies must sell and purchase operating assets that are more closely aligned with these new goals. We all have witnessed many mergers and acquisitions to achieve these objectives. In any case, these new operating facilities must be brought into the infrastructure of the acquiring enterprise. To do this without these standard building blocks (templates) is a daunting undertaking.

Effective Use of Analytics

As we write this book in early 2020, it seems the answer to every question in the world can be solved by advanced analytics and artificial intelligence (AI). While these technologies are powerful and solve many things, we have found that not all analytics are cloud-based or require AI. Our most successful customers have developed an analytics strategy that categorizes analytics from technical to business related and from simple operating calculations to complex and predictive analysis to integrated real-time business optimization. The best practice that we have found is to distribute the analytic as close to the edge, or to the manufacturing sensor, as reasonably possible, but no closer.

The reasons for this are speed of resolution, elimination of duplicating or transporting the data to different physical environments, and reduction of required labor. An EIDI system is quite capable of integrating time-series data and executing online analytics, whether diagnostic or descriptive. Some predictive analytics may be done within the EIDI if the EIDI has future time capability.

Complex modeling, rules-based analytics, data-based analytics, broad predictive analytics, and most business and financial analytics that integrate operations and production data should be performed with a specific time domain and scope relevant to the data and value. Often as the scope of data increases, the need to do this at the edge is lessened. The scope of the data (one or many sensors/sites) and the integration of multiple data sources will drive analytics to a more centralized configuration. These can be either on-premises or cloud-based. For those analytics, we recommend standard software utilities that extract, cleanse, contextualize, and publish data as input to analytics, machine learning, and analysis studios. Please refer to Chapter 10 for a more extensive description of this capability.

Self-Serve Access to Needed Information

The most successful enterprises set up work processes and software systems so that everyone needing the EIDI information has access to the information they need in the way they best consume it. Eliminating data silos and removing gatekeepers of information certainly help, but any EIDI system should be deployed in a manner that everyone can visualize, analyze, or configure what they need with a minimum of assistance. We have also seen a great many companies democratizing the information, allowing everyone from management to specialized knowledge workers, subject matter experts, operations, and maintenance personnel access to key information at all times, from any place via mobile devices. To do otherwise would be to take a fundamental tool out of the hands of the user. Imagine if spreadsheets needed to be created by an central IT-only function.

ProcIndustries Enterprise-Wide Rollout Results

The annual ProcIndustries' leadership meeting with Bill Roberts (vice president of operations) and Peter Argus (continuous-improvement manager) was scheduled for the upcoming week. They were to review progress made since ProcIndustries decided to proceed with the enterprise-wide EIDI deployment almost two years before. Since then, all of the refineries had deployed the EIDI and were using it as part of their daily work routine. By leveraging the South Texas refinery's methodology, configuration templates, and improved work processes, the remaining refineries were able to quickly use the EIDI data in an effective manner to reduce costs, optimize asset health, improve safety conditions, and resolve operational issues as they occurred. This was in part because of the hard work and vision of the original digital transformation team. But the personnel at the other refineries made it happen by their willingness to adapt and utilize what had been accomplished at the South Texas refinery to achieve the strategic goals described in Chapter 11.

In preparation for the meeting, Bill and Peter set aside a full day to strategize and prepare. Bill noted that all but one of the ProcIndustries refineries had met the target of returning the initial EIDI investment within two years. The one refinery that did not meet the goal had unexpected personnel turnover and were on track to reach their target in the next six months.

Cloud Strategy

Bill and Peter convened to prepare for the management meeting. Bill suggested that they start with the most pressing issue: defining cloud policy, implementation, data flow, and security for the next few years. Peter related that he had attended the latest OSIsoft conference and heard several companies discuss their specific approach. Peter also attended an oil and gas industry dinner during the conference, where he was able to informally discuss cloud strategy with his peers to and try to pick their brains on what actually constitutes best practices, if such things yet exist.

Peter found that most companies were grappling with several key issues: cloud design to facilitate more analytics offerings in the market, Internet of things (IoT) data integration with existing systems, and moving as many analytics as possible from the corporate environment to the edge, close to the refinery equipment to provide more timely value.

When Peter asked a fellow refiner, a strategic planning IT manager, about cloud design, she said that their company had embarked on a strategic initiative to architect their cloud to take advantage of the plethora of analytics coming onto the market that utilize production, operations, and equipment data. The first issue is always about security. Most operating companies, such as a downstream refining company, initially deployed a private cloud, where their data would be secure. However, this approach limited their ability to use newly available third-party packages that contain IoT sensors and analytics as single packages where data is sent to the public cloud.

The IT manager related that these new solutions are quickly installed and are very cost-effective, compared with the expense of connecting these sensors to existing control systems, both from the vendor cost, manual labor involved, and length of time to plan and implement such an initiative. She told Peter they had embarked on such a project to improve their vibration analysis on some of their rotating equipment. The vendor quickly installed the IoT sensors in one refinery and sent the results to their cloud analytics solution. The insights from this short-term project led to better operation of the units, and the maintenance team learned how to prevent abnormal situations.

She added that it was beneficial for them; however, after embarking on several of these successful projects, they had a lot of meaningful IoT data in individual vendor cloud environments. This resulted in siloed vendor clouds without a method of integrating the IoT data, analytics results, and vendor-added insights to their EIDI, which prevented many people being able to visualize and analyze the data in context to the rest of the process, production, and equipment data.

Because of this and many other business reasons, her company implemented a hybrid cloud, where much of the data is private but some of the data is public, mostly for analytics and public-facing information. Once that was decided, her company wrestled with how best to integrate the siloed analytics with their existing EIDI systems. She told Peter that after attending a cloud presentation at a conference, the answer might be a multi-tenant cloud environment, where different analytic and IoT sensor vendors might coexist in an environment where IoT data plus contextualized EIDI data can be easily and securely sent to this new cloud for analysis. At the same time, data and results from any of the analytic offerings can be integrated to the EIDI environment (and perhaps the legacy control systems) for further analysis, reporting, and collaborative notifications.

Peter took note of her key points and planned to talk to Pat Verlaine, ProcIndustries' IT director about how this could tie in to their existing cloud plans. Peter thought that if ProcIndustries could seamlessly extend the EIDI system to the cloud without needing to shift data to new systems, the time to value would decrease and the utility could easily increase.

Bill Roberts was intrigued by Peter's cloud conversation and asked Peter if he had any other takeaways from the conference. Peter shared that he learned a few interesting trends: The first was edge analytics and the hybrid cloud vision he saw and the second was the changing role of the process engineer at the modern plant or refinery. The two, he said, were intertwined.

Edge Analytics and the Evolution of the Control System

Thought leaders have stated that Industry 4.0 is about the effective use of digital information to improve all facets of manufacturing, including product quality, keeping assets in top condition, safety and health-related issues, environmental stewardship, optimizing production time, cost, and carbon impact of the product made. To achieve these goals, much of the improvement will be made from impactful use of existing data and complementing the existing data with new information needed for emerging analytics.

As analytics have proliferated and helped companies become more proficient and more agile, the big data analytics are usually performed at a corporate IT level. Some of that, especially when mixed with business data, must remain at that level. However, OT analytics is best when moved closer to the edge, where the manufacturing and equipment data is collected. This strategy is attractive for the following reasons:

1. The data does not have to go through an extraction and cleansing process, thereby reducing time to insight.
2. Fewer people in the company (IT, engineering, production, process engineering) need to be involved. Again, this reduces time and labor costs.

3. The strategy drastically reduces the time needed to make adjustments to the manufacturing process.
4. The edge analytic results are more easily integrated with EIDI systems, so that the quality of data is enriched by these results.

Peter told Bill that by bringing analytics closer to the edge, the roles of the process engineer and other local engineers at the refinery will change. They will be empowered to create these analytics using their EIDI system data to optimize the process governed by the traditional control system. Bill suggested that Peter look into hiring an engineering student or a recent graduate because engineering students at most universities are becoming much more well-versed in areas such as data science; AI; predictive tools, such as machine learning; and newer programming languages like Python and R. A recent graduate would have proficiency in machine learning and software analytics. This new worker could help develop solutions that would give the refineries greater insight on their equipment time-to-failure. The company could better prepare and budget to replace or avoid large or expensive capital equipment projects with smaller, more nimble projects.

Peter also had a vision of moving the analytics developer closer to the edge, where the developer would more quickly learn the refining process and what issues the refinery units face on a day-to-day basis. Bill digested what Peter related from his conference takeaways and remarked that this type of change will dramatically affect how control systems of the future will be designed and built.

Bill had attended ProcIndustries internal planning meetings where the company's top technical people were struggling with how to improve the existing refinery control systems and how to prepare and define what's needed for the next-generation control systems. Bill theorized that because of edge analytics, machine learning, and other breakthroughs in data modeling, the next wave of control systems will be quite different. For years, traditional control systems have been strictly OT based. That is, the fundamental components are process control logic, interlocks for safety, and data handling to efficiently make products.

When these systems are linked to modeling or planning software systems, they tend to use first-principles engineering models, OT equipment models based on vendor performance information and recommended service intervals, or engineering-based process models. They have not traditionally used modeling or predictive analytics that is, based on the data, for control. Nor have they used many of the IT analytics developed in the twenty-first century. They also don't natively adapt to adding IoT sensors in a multi-vendor environment without some data transmission work involved to integrate these sensor signals.

This is expected to change in the 2020s, however, when new control systems will be fundamentally changed to a more modular design that includes both traditional OT tools and IT analytics as well as the capability to ingest incrementally added IoT sensors. The OT landscape for control systems and supervisory control and data acquisition (SCADA) will change to incorporate much more IT specific functionality, such as data-dependent analytics, and become more flexible to business and consumer demand for dynamic production requirements.

At some point in the not too distant future, control systems will also be designed to function in lights-out manufacturing facilities without workers. As you would expect, this would start with facilities that do not have a high probability of malfunctions that would cause a fire, explosion, any kind of environmental incident, or harm the neighboring community. Certainly a refinery would be one of the last facilities, if ever, to do this. But it will likely follow the trend toward self-driving vehicles, where autonomous driving is implemented, controlled in stages, and is likely to be more safe than with humans driving.

The Leadership Meeting

That next week, Bill Roberts and Peter Argus met with ProcIndustries upper management, consisting of the business vice presidents, the chief information officer (CIO), and the Director of IT Pat Verlaine. Bill gave the team an update of the enterprise-wide EIDI rollout. He summarized how far each refinery had progressed in their EIDI adoption and the quantifiable benefits realized by each site.

All ProcIndustries refineries, except the last refinery to install their EIDI, had returns in excess of their total EIDI deployment expenditures. All refineries, except the one mentioned, recouped their initial EIDI investment within two years, although not exactly in the way it was originally justified.

The president of ProcIndustries thanked Bill and Peter for their hard work in making the company more competitive through use of digital technology and data analysis. He remarked, "You know Bill, when you first requested this, I wasn't sure we needed it at that time. We were prepared to upgrade some of our capital equipment at the refineries. But we now realize that by using the information correctly, we were able to increase capacity, reduce overall downtime by 5%, and reduce outages and maintenance costs significantly. We've also seen that the operations and business teams are working more closely together, not to mention the collaboration we've seen across the various refineries. But, between us, the EIDI deployment has mitigated many potential risks for us, especially unforeseen risks. By having actual historical data, we can produce any report we'd need and not be caught unprepared if we're a target of scrutiny or litigation. We can also show internal people and outside entities exactly how we've improved operations over time. All in all, well done, Now tell us what your plans for the next year are."

Peter spoke, discussing the upcoming initiatives that will be piloted at the South Texas refinery and will be deployed across the fleet, once successful testing is completed:

"We have three major initiatives we're planning this year: The first venture will be building on the successes we have had transitioning from reactive maintenance to a proactive model. We think we can accurately detect when equipment performance starts to deteriorate. When that happens, people get notified immediately and we can take corrective action before it becomes a real problem. The next step is to use the EIDI data more effectively by using analytics such as machine learning to predict when the equipment will fail or operate in a problematic state. This will allow us to effectively plan how to prioritize our maintenance spending and labor use to keep these assets running properly. We can also start forecasting when we need to replace faulty equipment."

Bill interjected: "We think there are enough potential savings here that, if we succeed, we will recommend to management that we create an internal center of excellence staffed by a team from the various refineries and corporate engineering to receive a feed of all equipment data and develop these predictive analytics in a standardized manner. This would alleviate local site people from having to worry about equipment health and run-to-failure problems. They would be notified only if there's an impending problem that affects near-term production. We think this pays for itself in less than 18 months."

Peter continued: "The second initiative is to provide hard hats that facilitate augmented reality. Which means that when personnel are out in the unit, they can see a digital representation of the current values and details of the equipment around them. This will help them identify problems much quicker and provide an additional layer of safety by displaying data that indicates there is a potential problem or unexplained equipment behavior.

"The third initiative is that we plan to share more information with our suppliers and customers securely through the cloud, so that our raw material suppliers and technology providers can better serve us by having an accurate view of what our inventory is in real time. This would enable us to receive supplies or services more quickly."

The leadership team thanked Bill and Peter, congratulating them on a job well done and looking forward to more information on their upcoming initiatives. Before they left, the president asked, "Bill, could you please reach out to each refinery and get the latest payback numbers on their EIDI rollouts? Please have it to me by the end of next week and be as thorough as possible. Thanks very much."

Hybrid Cloud Strategy

Bill and Peter debriefed immediately after the meeting. Both thought it went well but knew their vision and expertise were to be called on quickly in order to help IT finalize its cloud design in the next few weeks.

"What do you think, Peter?" Bill inquired.

"Bill, in my opinion, we need to make sure that we have the right environment for cloud analytics. This cloud structure needs to be designed so that our data science people can easily get our fleet-wide EIDI data and run their analytics. However, we also need a multi-tenant cloud so that we can take advantage of companies that can run analytics on our EIDI data and other data if we choose to engage them. I think we should really think through how many independent cloud instances we should be sending our data to," answered Peter.

"You're right." Bill interjected. "If we don't plan accordingly, we will be back to where we were 25 years ago where the refinery data were in different databases that didn't communicate with each other."

Peter hoped that their design would facilitate a seamless transfer of data and a smooth workflow between the refinery and their edge processing, the enterprise, and their emerging cloud environment. He told Bill, "What we should investigate is a strategy of not replicating our EIDI data to an inordinate number of vendor-supplied independent cloud instances. We should carefully target where we replicate our data. If our EIDI vendor could facilitate a multiple vendor analytics community environment in the cloud, this would be more straightforward and easier to manage from a data replication standpoint."

He knew that more cloud analytics would spawn a multitude of more complex analytic offerings where they could leverage their contextualized EIDI data. Peter hoped they would soon reach the point where more technologically advanced big data analytics would be available for them to pursue.

Peter told Bill, "Let's start planning for the community environment. I've attended several cloud conferences and I think a big issue on the horizon is being able to ingest and serve out EIDI and business data simultaneously to multiple entities. They may be suppliers, customers, service providers, the city, who knows. Think of it as a superset of what we've already done with the cloud-based connected services." (See Chapter 9.)

"Pretty soon we will be able to automatically do business with our raw material suppliers through the cloud, where they have access to our inventories and will supply product to us when they detect in real time that we need restocking."

Peter and Bill decided this needed quick attention and set a meeting in two weeks with IT Director Pat Verlaine.

When they met, they started to share their cloud vision with Pat. Pat listened intently but seemed a bit distracted. When Bill asked, "Pat, are we making sense here?" Pat replied that it all seemed reasonable, but priorities might be changed. Peter asked why. Pat disclosed that there were rumors of an imminent ProcIndustries acquisition.

Pat continued, "I'm not at liberty to say much, but I was brought in to evaluate another company's suite of IT and OT products to determine compatibility between the two companies."

Bill asked Pat, "Are we getting bought out?"

"No," Pat said. "Actually, we are looking to acquire a number of petrochemical plants from a well-known chemical company, who I can't reveal. SEC rules, you know, and please keep this conversation between us. The chemical company wants to shift its focus to specialty products and precision agriculture, you know, genetically engineered food. I don't know, maybe another alternative meat company. In any case, they want out of the basic chemicals market and we apparently want in to add to our downstream value chain."

Pat went on, "It seems that all of the cost reductions and increased revenue from the company-wide EIDI deployment has translated into a tidy sum for ProcIndustries, thanks to you two."

Bill countered, "Thanks very much, Pat. But you know we couldn't have done it without you personally and your entire IT staff, especially with the big data analytics and the cloud-based connected services. You deserve credit as much as we do. I hate to run off, but thanks for the heads-up. We'll keep what you just told us quiet. Right, Peter?"

The Next Challenge

Pat Verlaine was correct. Within weeks of the discussion, the acquisition was announced. ProcIndustries would be purchasing three additional olefins petrochemical sites, each of which included an ethylene plant, propylene plant, and benzene plant. The move was made to take advantage of emerging market growth in Mexico and Central America.

Ten days after the announcement, Peter, Bill, and Pat attended a meeting with the ProcIndustries leadership team. The president thanked them for being available to discuss the upcoming acquisition.

He continued, "Bill, Peter, thanks again for the wonderful job you've done transforming our refineries. We're now in very good shape. As such, we'd like you two and Pat to form a new team and transform these new petrochemical plants the way you did our refineries. From what I hear, they weren't run in a very effective manner. They don't have anywhere near the same comprehensive sets of data that we now have here. We understand that the plants we are buying don't even communicate effectively among themselves, as their control and information systems are all from entirely different vendors and they've never integrated them effectively."

"That's right," Pat chimed in. "It's a hodgepodge of vendors and systems. And very limited uses of analytics in a systematic way."

"We know exactly what to do," exclaimed Peter. "Let's get a meeting set to review together and get started. It will be like old times."

With that, they soon convened and started a new journey. They underwent the same methodology and took the same steps they had taken when improving the refineries. This time, they knew exactly what steps to take and what things they should not do. In less than a year, they had an effective data infrastructure for these newly acquired plants. They deployed an EIDI system at every site, transmitting a subset of the data to a petrochemical division-wide EIDI that focused on business systems integration, advanced analytics, modeling, and scheduling.

They also had accomplished new initiatives:

- They integrated all of the equipment performance information directly into the newly deployed ProcIndustries Center of Excellence operations monitoring center, staffed with domain specialists to monitor and prioritize potential equipment issues. This allowed plant personnel to focus on process monitoring and meeting production targets.
- They deployed geographic-based environmental and safety condition software that monitors environmental impact from the petrochemical plants and transmits real-time EIDI emissions values to the geospatial information system. By superimposing EIDI real-time values on these maps, operations people would have real-time situational awareness that could impact safety decisions. Then, based on historical emissions data, they can predict if the site would reach an emissions limit governed by the U.S. Environmental Protection Agency (EPA) and the State of Texas.

The team also made plans to add offline process modeling of the petrochemical processes. They planned to send predicted values from the software model back to the petrochemical site-based EIDI systems and store these predictions as EIDI future data values. By doing this, operations personnel could track the actual process results versus what the process model had predicted and take action when a significant variance occurred.

And the ProcIndustries' Story Continues

Approximately one month after completing the preceding projects for the petrochemical plants, Bill Roberts and Peter Argus were called into the president's office one more time. When they met, the president again thanked them for integrating the newly acquired petrochemical plants. He told them the margins were starting to improve and having access to the data certainly helped that effort. Then he inquired, "How about another challenge?" Bill asked him if ProcIndustries was shopping for more capacity to add to the existing fleet. The president replied: "No, Bill. I think we are out of the acquisition business, at least for now. But I have another idea."

He continued, "One of my responsibilities is always looking for new sources of revenue and profitability to grow our business. Sometimes that entails acquisitions; sometimes it involves plant debottlenecking or other large cost-cutting initiatives. But we're thinking about something new here. You and your teams have done such a wonderful job transforming our company. Now we are a data-driven enterprise and operate much more efficiently. Our profitability certainly reflects that." He added, "Therefore, I'd like you two to consider starting a new ProcIndustries business, reporting directly to me. We'll call it the *Digital Transformation Group*."

Peter inquired, "What exactly would we do? Keep the company current with technology?"

The president answered, "Yes, that would certainly be part of the scope. But as I see it, the main thrust would be to provide thought leadership to other companies that are struggling with digital transformation." He added, "I meet with many other presidents and CIOs in our business. The smaller to mid-size companies need some guidance. So I thought we could help, as a revenue arm of the company."

"We would show them the correct way to identify and deploy technology effectively. We would be able to advise them on several things, such as how to correctly align their workforce to collaborate and improve workflow processes. Since we know the refining and chemical plant business from the ground up, we would be credible from day one."

Bill and Peter exclaimed, "Sounds great! When do we start?"

The president replied, "Great! This will be an enjoyable assignment for both of you. Let's start laying out a plan next week."

As they left the president's office, Peter told Bill, "Well, I guess we are finally digitally transformed."

What You Should Take Away

A business transformation, enabled by improved digital technology, vision, hard work, and improved workflows can make a tremendous improvement to a company's financial health, the physical health and safety of its communities, the relationships with its customers and suppliers, its reputation as a thought leader in its industry, and the morale of its employees by being on a winning team.

Throughout the book, we have identified and described how to implement the key use cases that will turn your data into an extremely valuable resource, and as a result, strengthen companies. In Chapter 8, we summarized five companies that have used EIDI technology in the most successful and innovative ways. We hope you take something from these examples.

In this chapter, we shared a holistic blueprint for success by articulating the most important decisions and the best actions that the most successful companies have taken to achieve operating excellence and hasten their transformational journey.

We also presented topics that many companies currently face, particularly companies that have succeeded with enterprise-wide EIDI deployments. Having triumphed in reducing costs, improving their processes, and keeping their equipment in peak operating condition, they must now make wise decisions in cloud planning and deployment, using IoT devices to augment their existing information, pursuing autonomous operation of plants and equipment, and moving information beyond their firewalls and protected domains into an emerging community-based IT environment.

Finally, you have taken the journey with the ProcIndustries digital transformation team and have seen them transition from a reactive business environment to a data-driven company that continually tightened up inefficiencies and put their company in a position where they could acquire more assets to strengthen their business even more.

> Thank you for reading and best of luck in your own journey to digital transformation!

Additional Reading

Harclerode, C. 2019. "Five best practices of digital transformation initiatives in oil and gas." White paper, OSIsoft. https://explore.osisoft.com/wp-5-best-practices-dx-og/whitepaper-five-best-practices?utm_campaign=wp-syndication-hydrocarbon-processing&utm_medium=referral&utm_source=hydrocarbon-processing.

Index

Note: Page numbers in italic and bold refer to figures and tables, respectively.

A

AAP, see Anglo American Platinum (AAP)
abnormal conditions and trigger events, 143
advanced analytics integration, 235
  and ML, 187–188
  production event-based data to, 235–239, *237*
  types, 227–228, *236*
  use of, 268–269
advanced pattern recognition (APR), 138, 150
advanced visual analytics, 96–100
AF, see asset framework (AF)
AI, see artificial intelligence (AI)
Alex's EIDI spreadsheet report, *176*, **176**
Anglo American Platinum (AAP), 217–219
  asset data hierarchy, *219*
  benefits for, 219–220
  PI CC environment at, *218*
anomaly detection, 138
APR, see advanced pattern recognition (APR)
ArcGIS, 240
artificial intelligence (AI), 268
  and closed-loop control, 204–205
  Cortana, *99*
asset framework (AF)
  database, 233
  model, 268
  object, 72
  PI AF, see PI Asset Framework (PI AF)
asset optimization, 135
asset relative displays, 121–122, *122*
assets and event frames, 73–75

B

BI, see business intelligence (BI)
big data analytics, 198, 229, *229*, *232*
block flow diagrams, **78**, 78–79
business digital transformation, *33*
business intelligence (BI), 115
business objectives, 228–230
  pyramid, 130
business processes workflows diagram, *42*

C

CBM, see condition-based maintenance (CBM)
cement manufacturing company, *201*
centrifugal pump system, *147*
ChemCo, 184, 205
  fleet-wide CBM program, 207
  on improving operations, 206
  three-tiered energy management system, 207
  using technology, 205–206
cloud-based BI, 115
cloud-based services, 220–222
cloud strategy, 280–281
competence centers, 54–55
condition-based maintenance (CBM), 137, 140, *142*, 143
  fleet-wide program, 207
  implementations, 149
  scenarios, 145
condition-based monitoring, 133
control system evolution, 281–283
coordination controls process, 132
corporate leadership and plant employees, 33–34
Cortana AI, *99*

281

crude unit
  production and consumable losses, 95
  relative dashboard, 94
customer success manager (CSM), 256
  responsibilities, 256, 256–258
  role of, 257

D

data
  acquisition/validation/classification, 77–78
  analysis/visualization/reporting, 169–173
  capture and reporting, 169
  -driven analytics, 100–103
  lake approach, 233
  management, 65–66
  object models, 75–77
  reconciliation system, 178–179, 179
  visibility, 67
DCSs, *see* distributed control systems (DCSs)
decentralized decisions, 27
De Geus, A., 2–3
demand/response management, 167
demilitarized zone (DMZ) strategy, 260
derived/inferred variables, 66
digital data infrastructure, 6–7, 50–51
digital enterprise
  business goals with, 266
  challenge, 286–287
  edge analytics and control system evolution, 281–283
  operations/production data as strategic asset, 267
  self-serve access, 269
  strategic partners, 267
  unexpected, architecting for, 268
digital plant template, 87, 89, 236
digital twin concept, 239–240
distributed control system (DCS), 7, 12
DMZ, *see* demilitarized zone (DMZ) strategy
downtime cost, 23
dynamic performance operational displays, 124–126

E

edge analytics, 281–283
EIDI, *see* enterprise industrial data infrastructure (EIDI)
EIDI Asset Framework (AF), 233–234
energy consumption, 163–165
  business objectives, 165–167
  ISO 50001 standard, 165
  real-time visibility to, 251
  and water, 171, 172
energy management, 163
  event, 167
  power consumption, 173
  process flow diagrams for, 175–177
  refinery's energy demand/consumption, 167–168
engineering group's role, 22
enhancing equipment availability, 135–136
  CBM, 137
  condition-based maintenance, 139–141
  predictive maintenance, 137–138
  preventive maintenance, 136
  reactive maintenance, 136
enhancing process control performance monitoring, 154–157
enterprise asset management, 135
enterprise competence center, 52–54
enterprise data architecture, 53
enterprise industrial data infrastructure (EIDI), 2–4, 23, 65–68
  architecture and cybersecurity, 260
  deployment and configuring, 73–75
  energy consumption, 164
  enterprise operational and BI, 115
  enterprise rollout team, 255–256
  functionality/data ingress/egress, 68
  implementation approach, 69–71
  innovative use of, 88–103
  need for, 61–63
  people-driven benefits from, 116
  real-time notifications/work orders, 152
  system, 179
  value over time *versus* value limited, 261
enterprise level, 54

Index

enterprise-wide EIDI
   deployments, workflow for, 259, *259*
   implementation, 254, *255*
environmental regulations, 5
equipment parameter tables, 144–145
equipment template, assigning context, 142–143
ERP (enterprise resource planning), 8, 26, 47, 52
Esri ArcGIS, 241, *241*
event frame data, **91**, *75*
event management, 133–134
expert and advanced users, 111–112

F

fouling, 5

G

Gantt chart projecting, *260*
generalized continuous improvement, *52*
geospatial systems, time-series data with, 240–241
GoldMineCo, 184, 195
   big data analytics, 198
   digital plant template, 199
   mines and plants, 197
   for mobile asset maintenance, 197–198
   plans, 199–200

H

heat exchanger fouling, 176, *176*
hierarchical metadata structure, 185–186
hybrid cloud strategy, 285–286

I

ICC, *see* Integrated Collaboration Center (ICC)
idle state management, 167
IIoT, *see* industrial Internet of things (IIoT)
impact changes, EIDI system, 112–113
improving asset availability, implications, 152–154
individual refineries, 55–56

industrial digital data infrastructure
   current, 9–12
   current state, assessing, 4–5
   disaster, 1–2
   EIDI, 2–4
   problem identifying, 6–9
industrial Internet of things (IIoT), 54, 138, 222
industrial maintenance strategies, *135*
Industry 4.0, 27
information gathering, 10
information transparency, 27
infrastructure, 6
   software supplier, 63–64
Integrated Collaboration Center (ICC), 191
integrated oil and gas company, *188*
integrated ProcIndustries refinery, *168*
integration with systems, 151–152
International energy standards, 165
interoperability, 27
ISO 50001 standard, 165
IT department's role, 21–23

J

jump-start workshops, 259
"just-in-time" model, 216

L

laboratory information management systems (LIMS), 8
layers of analytics strategy, 185
leadership meeting, 283–284
left side, business objectives pyramid, 131–133
LIMS, *see* laboratory information management systems (LIMS)
linear optimization modeling, 238
linear programing (LP), 17, 194, 238
"the living company," 2
long-term value initiatives, *114*
LP, *see* linear programing (LP)

M

machine learning (ML), 66, 231, 242
maintenance manager's view, 20–21
management's decision, 253–254

management users, 111
mass balances and data reconciliation, 178–179, *179*
MaterialsCo, 184, 200
    corporate BI software, 202–203
    initiatives, 200–201
    PI System software, 201
    production management version 2.0, 203
MATLAB and software languages, 227–228
maturity state, 27
metering and inputs, 168
Microsoft Azure, 187
Microsoft Excel, 227
    analytics tools, 100
    Power Pivot, *172*
Microsoft Power BI, *238*
    desktop, *97*, *98*
MidPetCo, 183, 190
    Energy Lab, 191
    initial use cases, 192
    overview and challenges, 191
    real-time visibility, 193–194
    return on investment and plans, 194–195
MidPetCo 2.0, 191
    objectives of, 194–195
    profitability, operational data for, 191–192
mining operations, 218
ML, *see* machine learning (ML)
mobile access, information, 120–121, *121*
multivariable statistical process control charts (MSPCs), 177
multivariate controls, 132

N

Natural Resource Canada's energy management methodology, 165, *166*
New Downstream Program (NDP), 184
non-quantifiable benefits, 252–253

O

ODBC, *see* open database connectivity (ODBC)
offline and big data analytics, 228
OilCo, 183–184
    advanced analytics and ML, 187–188
    analytics framework, 185
    EIDI deployment, 185–186
    fundamental strategies, 184
    return on investment, 189
oil refinery block diagram, *90*
online EIDI analytics, 227
on-premise *versus* cloud analytics, 188–189
open database connectivity (ODBC), 96
operating envelopes, 119
operating modes, *170*, *171*
operational excellence, 6
    program, 38
operational excellence methods and tools, 46
    first loop, 48–49
    second loop, 49–50
    workflow management, 47–48
operational expenditures (OPEX), 88, 251
operational intelligence to enterprise intelligence, 226
operational performance management, 154–157
operational sensors, 45
operational team
    convening, 31–32
    importance, 33
operational variance algorithm, 100
OPEX, *see* operational expenditures (OPEX)
organizing operational data, 81
OSIsoft PI System, 214
Outotec, 217
    AAP, 217–219, *218*, *219*
    benefits for, 220
    remote service methodology and process workflow, *220*

P

PCA, *see* principal component analysis (PCA)
peak demand management, 167
people-centric loop, 49
performance monitoring, 134
PFCDs, *see* process flow control diagrams (PFCDs)

Index

P–F curve, *139*, 139–141
PI Asset Framework (PI AF), 185–186
  templates, *193*
PI Cloud Connect (PI CC)
  architecture, *216*
  customer examples, 217–219
  overview, *217*, 222–223
  scalability, *222*, 222–223
PIMS, *see* Plant Information Management System (PIMS)
pivotal component, *188*
planning and economics coordinator's view, 16–17
planning and execution, 134
plant block diagrams, 89–93
plant data hierarchy, decision-making, 71–72
Plant Information Management System (PIMS), 200
plant manager, 1–2, 48
  view, 15–16
plant systems, **13–14**
PLCs, *see* programmable logic controllers (PLCs)
PLS, *see* projection of latent squares (PLS)
Power BI
  energy, *173*
  report template, 236
power generation company, 220–222
principal component analysis (PCA), 177
process control
  engineers, 22
  and operational intelligence, *131*
process engineers, 22, 49
  view, 12–15
process/equipment models and simulations, 228
  digital twin for, 239–240
process flow control diagrams (PFCDs), 15
process flow diagrams, 79–81
process improvement, asset behavior for, 96
process manufacturing company, *118*
process unit, 2, *74*
process unit templates, 44, 199
  smart thinking, 44–46

ProcIndustries, xix–xxi, 1, *10*, 29, *76*
  current and target states, **29**–**30**
  EIDI users, 111–112
  energy process diagram, *169*
  integrates operational and business data, 230–232
  physical model, **82**
  story continues, 288
ProcIndustries enterprise-wide rollout, 245
  business case, 247
  digital transformation, 245–247
  EIDI, 256–258
  future use cases, 260–261
  initial use cases, 255–260
  justifying proposed EIDI, 248–249
  management's decision, 253–254
  results, 269
production manager's view, 17–19
programmable logic controllers (PLCs), 7
projection of latent squares (PLS), 177
project leader importance, 32
pump asset example, 146–148
pump characteristics and variables, **149**
pump monitoring and analysis, 148–150

Q

quantifiable benefits, 250–252

R

real-time continuous improvement, *47*
real-time data analytics, 130
  business objectives pyramid, 131–134
real-time data infrastructure, 141–142
  versus data lake approach, 233–234
real-time EIDI data infrastructure, *43*
real-time energy monitoring, 167
real-time infrastructure, *63*
real-time notification ability, *95*
refinery equipment, 163
refinery manager's office, 37–39
refinery overview dashboard, *93*
refining and petrochemical applications, *187*
regional level, 54–55

regulatory controls, 131–132
right side, business objectives pyramid, 133–134
root cause determination, 144

S

SCADA, *see* supervisory control and data acquisition (SCADA)
self-serve access, 269
simplified control monitor, **156**
Six Sigma methodology, 40
smart grid and refinery resiliency improvements, 173–175
smart process unit template
    components, *92, 93*
    schematic, *51*
software data infrastructure, 6
South Texas refinery process block diagram, *79*
statistical process control (SPC), 122
statistical quality control (SQC) charts, 227
strategy envisioning, 24–25
    assessing data infrastructure maturity, 26–29
    breakthrough vision, 25–26
    foundation laying, 31–34
subject matter experts and technical users, 111
supervisory control and data acquisition (SCADA), 13
supervisory process control systems, 8
supply chain optimization, 132–133

T

target and improvement loops, **50**
technical assistance, 27
three-tiered energy management system, 207
time-series data, 45, 226
time-series database, 12
traditional *versus* remote troubleshooting scenarios, *221*

U

unexpected, architecting for, 268
unit process templates, 89–93

V

visualization methodology, 117–118
visualization tools types, *119*, 120
visualizing operational variance, 93–95
visualizing pump actionable output, 150–151

W

water management, 197
workflow information gaps, identifying, 40–41

Made in the USA
Monee, IL
03 May 2026

49437440R00177